ADVANCES IN MATERIALS RESEARCH 3

Springer
*Berlin
Heidelberg
New York
Barcelona
Hong Kong
London
Milan
Paris
Singapore
Tokyo*

Physics and Astronomy

http://www.springer.de/phys/

ADVANCES IN MATERIALS RESEARCH

Series Chief Editor: Y. Kawazoe

Series Editors: M. Hasegawa   A. Inoue   N. Kobayashi   T. Sakurai   L. Wille

The series Advances in Materials Research reports in a systematic and comprehensive way on the latest progress in basic materials sciences. It contains both theoretically and experimentally oriented texts written by leading experts in the field. Advances in Materials Research is a continuation of the series Research Institute of Tohoku University (RITU).

1   **Mesoscopic Dynamics of Fracture**
    Computational Materials Design
    Editors: H. Kitagawa, T. Aihara, Jr., and Y. Kawazoe

2   **Advances in Scanning Probe Microscopy**
    Editors: T. Sakurai and Y. Watanabe

3   **Amorphous and Nanocrystalline Materials**
    Preparation, Properties, and Applications
    Editors: A. Inoue and K. Hashimoto

Series homepage – http://www.springer.de/phys/books/amr/

A. Inoue   K. Hashimoto (Eds.)

Amorphous and Nanocrystalline Materials

Preparation, Properties, and Applications

With 169 Figures

Springer

Professor Akihisa Inoue
Professor Koji Hashimoto
Institute of Materials Research, Tohoku University
2-1-1 Katahira, Aoba-ku, Sendai 980-8577, Japan

Series Chief Editor:

Professor Yoshiyuki Kawazoe
Institute of Materials Research, Tohoku University
2-1-1 Katahira, Aoba-ku, Sendai 980-8577, Japan

Series Editors:

Professor Masayuki Hasegawa
Professor Akihisa Inoue
Professor Norio Kobayashi
Professor Toshio Sakurai
Institute of Materials Research, Tohoku University
2-1-1 Katahira, Aoba-ku, Sendai 980-8577, Japan

Professor Luc Wille
Department of Physics, Florida Atlantic University
777 Glades Road, Boca Raton, FL 33431, USA

ISSN 1435-1889
ISBN 3-540-67271-0 Springer-Verlag Berlin Heidelberg New York

Library of Congress Cataloging-in-Publication Data applied for.

Die Deutsche Bibliothek - CIP-Einheitsaufnahme

Amorphous and nanocrystalline materials : preparation, properties, and applications / A. Inoue; K. Hashimoto (ed.). – Berlin; Heidelberg; New York; Barcelona; Hong Kong; London; Milan; Paris; Singapore; Tokyo: Springer, 2001
(Advances in materials research; 3)
ISBN 3-540-67271-0

This work is subject to copyright. All rights are reserved, whether the whole or part of the material is concerned, specifically the rights of translation, reprinting, reuse of illustrations, recitation, broadcasting, reproduction on microfilm or in any other way, and storage in data banks. Duplication of this publication or parts thereof is permitted only under the provisions of the German Copyright Law of September 9, 1965, in its current version, and permission for use must always be obtained from Springer-Verlag. Violations are liable for prosecution under the German Copyright Law.

Springer-Verlag Berlin Heidelberg New York
a member of BertelsmannSpringer Science+Business Media GmbH

© Springer-Verlag Berlin Heidelberg 2001
Printed in Germany

The use of general descriptive names, registered names, trademarks, etc. in this publication does not imply, even in the absence of a specific statement, that such names are exempt from the relevant protective laws and regulations and therefore free for general use.

Typesetting: Dataconversion by LE-TEX, Leipzig
Cover concept: eStudio Calamar Steinen
Cover production: *design & production* GmbH, Heidelberg

Printed on acid-free paper   SPIN: 10732219   57/3141/tr - 5 4 3 2 1 0

Preface

Amorphous and nanocrystalline alloys are relatively new functionally important materials that demonstrate superior properties in a wide range of conditions and have a number of important applications. They include materials with exclusively good static and dynamic mechanical strength, magnetic properties, good corrosion and wear resistance. Some of them exhibit superplasticity including high strain rate flowability. It was also shown that precipitation of nanocrystalline particles from the amorphous matrix improves the tensile strength and ductility of the alloy. Nanocrystalline alloys consist of particles/grains with sizes ranging from 1 to 100 nm having common boundaries or embedded in the amorphous matrix-forming nanocomposite. High-strength nanocrystalline alloys obtained in Al-, Mg- and Zr-based systems attract significant commercial interest due to their high strength. Amorphous and nanocrystalline materials can be produced depending on whether the crystal nucleation frequency is extremely small or large, respectively, at sufficiently low growth rates. The conditions that allow formation of such materials depend on the liquidus, glass transition and crystallization temperatures, the thermodynamic quantities such as the crystal-interfacial energy and the free energy driving force for homogeneous versus heterogeneous crystal nucleation, the viscosity of the supercooled liquid and its temperature dependence, the liquid alloy packing density and the volume change during crystallization.

This book contains seven chapters. They cover a wide area of metastable amorphous and nanocrystalline materials including nanomartensite. Special chapters are devoted to the structural relaxation, diffusion and electrical resistance behaviour of amorphous alloys. Well-established formation and processing methods for amorphous and nanocrystalline alloys, including mechanical milling, electrodeposition and sputter deposition, are discussed, together with the properties of the materials produced. Metalloid-based amorphous and nanostructured materials produced by rapid solidification of the melt are also described in a chapter. It is also shown that bulk samples of such materials can be produced by hot-pressing. A chapter describing novel materials for CO_2 recycling used to prevent global warming is also included. Formation of nanomartensite grains in steel and its role in fatigue strengthening of steel are described in the last chapter. The first chapter is devoted to bulk amorphous alloys, including their production, properties and applications. Bulk amorphous alloys are relatively new, and represent a quickly developing field of materials science. As is known, high cooling rates above 10^5 K/s are usually required to produce metallic glasses, which restricts the size of the samples to about several tens of micrometers and limits their applications. However, in some particular systems bulk amorphous alloys with sizes of more than 1 mm were obtained. Zr-, Ti-, Fe- and Mg-based amorphous alloys are the most important ones from a commercial viewpoint. Mechanisms for achieving high glass-forming ability and various properties of bulk amorphous alloys are discussed.

The increasing publication pressure in the field of bulk amorphous and nanocrystalline alloys, as well as the large number of international meetings and conferences organized (for example: the 4th and 5th International Conferences on Nanostructured Materials NANO 98 and Nano 2000, recently held in Stockholm, Sweden and Sendai, Japan, respectively; Supercooled Liquid, Bulk Glassy, and Nanocrystalline States of Alloys Symposium, to be held during the 2000 Fall Meeting of the Materials Research Society in Boston, USA; Bulk Metallic Glasses Conference in Singapore, September 2000; and many other meetings related to the subject) reflect the significant interest of scientists all over the World in this field of materials science.

This book shows the progress achieved in various areas of amorphous and nanostructured materials and is expected to be of interest to scientists working in the field of non-equilibrium materials, and of benefit to students.

Sendai,
October 2000

A. Inoue
K. Hashimoto

Contents

1 **Bulk Amorphous Alloys**
A. Inoue 1
1.1 History of Bulk Amorphous Alloys 1
1.2 Dominant Factors for High Glass-Forming Ability 3
1.3 Crystal Nucleation and Growth Behavior
    of Alloys with High GFA 8
1.4 Continuous Cooling Transformation
    of Alloys with High GFA 10
1.5 Preparative Methods and Maximum Thickness
    of Bulk Amorphous Alloys 11
1.6 Structural Relaxation and Glass Transition 15
1.7 Physical Properties 20
    1.7.1 Density 20
    1.7.2 Electrical Resistivity 20
    1.7.3 Thermal Expansion Coefficient 23
1.8 Mechanical Properties 24
1.9 Viscoelasticity 29
1.10 Soft Magnetic Properties 34
    1.10.1 Formation and Soft Magnetic Properties
           of Bulk Amorphous Alloys 34
    1.10.2 Glass-Forming Ability of Fe-(Al,Ga)-Metalloid,
           Fe-TM-B, and Co-TM-B Alloys 38
1.11 Viscous Flow and Microformability
     of Supercooled Liquids 39
    1.11.1 Phase Transition of Bulk Amorphous Alloys 39
    1.11.2 Deformation Behavior of Supercooled Liquids 40
    1.11.3 Microforming of Supercooled Liquids 41
1.12 Bulk Amorphous Alloys Produced
     by Powder Consolidation 43
    1.12.1 Consolidation Conditions 43
    1.12.2 Density and Properties
           of Consolidated Bulk Amorphous Alloys 44
1.13 Applications and Future Prospects 47
References 48

2 Stress Relaxation and Diffusion in Zr-Based Metallic Glasses Having Wide Supercooled Liquid Regions
Y. Kawamura, T. Shibata, A. Inoue, T. Masumoto, K. Nonaka, H. Nakajima, and T. Zhang 52

2.1 Introduction 52
2.2 Experiments 53
2.3 Results and Discussion 54
    2.3.1 Stress Relaxation in $Zr_{65}Al_{10}Ni_{10}Cu_{15}$ Metallic Glass 54
    2.3.2 Diffusion in $Zr_{55}Al_{10}Ni_{10}Cu_{25}$ Metallic Glass 61
2.4 Conclusions.. 67
References 67

3 The Anomalous Behavior of Electrical Resistance for Some Metallic Glasses Examined in Several Gas Atmospheres or in a Vacuum
O. Haruyama, H. Kimura, N. Nishiyama, T. Aoki, and A. Inoue 69

3.1 Introduction 69
3.2 Experimental Procedure 71
3.3 Results and Discussion 71
    3.3.1 Pd-Si Based Glasses 71
    3.3.2 $Pd_{40}Ni_{10}Cu_{30}P_{20}$ Glass 77
    3.3.3 $Zr_{60}Al_{15}Ni_{25}$ Glass 79
    3.3.4 Change in Electrical Resistivity Associated with Glass Transition 80
3.4 Concluding Remarks... 84
References 85

4 Methods for Production of Amorphous and Nanocrystalline Materials and Their Unique Properties
T. Aihara, E. Akiyama, K. Aoki, M. Sherif El-Eskandarany, H. Habazaki, K. Hashimoto, A. Kawashima, M. Naka, Y. Ogino, K. Shimiyama, K. Suzuki, and T. Yamasaki 87

4.1 Introduction 87
4.2 Crystalline-Amorphous Cyclic Transformation of Ball Milled $Co_{75}Ti_{25}$ Alloy Powder 88
    4.2.1 Use of Mechanical Alloying Technique for Amorphization . 88
    4.2.2 Ball Milling Procedure and Analyzing Technique 88
    4.2.3 Structural Changes vs. Milling Time 89
    4.2.4 TEM Observations 90
    4.2.5 Magnetization 92
    4.2.6 Thermal Stability 93
    4.2.7 Possible Reasons for the Cyclic Crystalline-Amorphous Transformations 95

4.3	Formation of Amorphous and Nanocrystalline Ni-W Alloys by Electrodeposition and Their Mechanical Properties		96
	4.3.1	Electrodeposition – A Method for the Production of the Amorphous Materials	96
	4.3.2	Preparation of Ni-W Alloys and Technique Used for Studies	97
	4.3.3	Brittleness of the As-electrodeposited Ni-W Alloys	104
	4.3.4	Hardness of the Nanocrystalline Ni-W Alloys	105
4.4	Formation of Ti-Based Amorphous Alloys by Sputtering and Their Physical Properties		109
	4.4.1	Sputtering Technique	109
	4.4.2	Samples Preparation and Description of the Analytical Equipment Used	109
	4.4.3	Structure and Mechanical Properties of Sputtered Alloys	110
	4.4.4	Amorphous to Crystalline Phase Transition	114
4.5	Hydrogen Evolution Characteristics of Ni-Mo Alloy Electrodes Prepared by Mechanical Milling and Sputter Deposition		115
	4.5.1	CO_2 Recycling Problem	115
	4.5.2	Experimental Procedure	116
	4.5.3	Mechanically Alloyed Ni-Mo Electrodes	118
	4.5.4	Sputter-deposited Ni-Mo Electrode	122
4.6	Concluding Remarks		128
References			130

5 Amorphous and Partially Crystalline Alloys Produced by Rapid Solidification of The Melt in Multicomponent (Si,Ge)-Al-Transition Metals Systems
D. V. Louzguine and A. Inoue 133

5.1	Introduction		133
5.2	Multicomponent Fully Amorphous Si and Ge-based Alloys		135
	5.2.1	Influence of Composition and Cooling Rate on the Structure of (Si,Ge)-Al-TM Alloys	135
	5.2.2	Reasons for the Elevated Glass-forming Ability	139
	5.2.3	Properties	140
	5.2.4	Thermal Stability and Crystallization of the Amorphous Phase	142
	5.2.5	Production of Bulk Amorphous Samples by Hot Pressing. Densification Behaviour.	146
5.3	Precipitation of Nanocrystalline c-Ge Particles in Mixed Si-Ge-Al-TM and Ge-Si-Al-TM Alloys		150
	5.3.1	Microstructure and Phase Composition of Rapidly Solidified Si-Ge-Al-TM Alloys	150

		5.3.2	Crystallization Process in the Rapidly Solidified Si-Ge-Al-TM Alloys	157

     5.3.2 Crystallization Process
            in the Rapidly Solidified Si-Ge-Al-TM Alloys 157
     5.3.3 The Effect of Si Addition to Melt Spun Ge-Al-TM Alloys . 160
References 164

6 Global CO_2 Recycling – Novel Materials, Reduction of CO_2 Emissions, and Prospects
K. Hashimoto, K. Izumiya, K. Fujimura, M. Yamasaki, E. Akiyama, H. Habazaki, A. Kawashima, K. Asmi, K. Shimamura, and N. Kumagai 166

6.1 Introduction 166
6.2 Global CO_2 Recycling 167
6.3 Key Materials for Global CO_2 Recycling 169
    6.3.1 Cathode Materials 169
    6.3.2 Anode Materials 174
    6.3.3 Catalysts for CO_2 Methanation 179
6.4 A Global CO_2 Recycling Plant for Substantiation of the Idea 182
6.5 Energy Balance and Amounts of Reduction of CO_2 Emissions 183
6.6 Economy of the Global CO_2 Reduction 184
6.7 Concluding Remarks 185
References 185

7 Formation of Nano-sized Martensite and its Application to Fatigue Strengthening
M. Shimojo and Y. Higo 186

7.1 Introduction 186
7.2 Formation of Micro-sized Martensite 186
7.3 Formation of Nano-sized Martensite 191
7.4 Application of Micro and Nano-sized Martensite to Materials Strengthening 196
7.5 Conclusions and Future Work 204
References 204

Index 205

1 Bulk Amorphous Alloys

A. Inoue

Institute for Materials Research, Tohoku University, Sendai 980-8577, Japan

Summary. This chapter aims to review our recent research results on new bulk amorphous alloys. The main topics are the following: (1) the finding of new amorphous alloys with high glass-forming ability in a number of alloy systems; (2) the mechanism for achieving high glass-forming ability; (3) the fundamental properties of the new amorphous alloys; (4) successful examples of producing bulk amorphous alloys by different techniques of water quenching, metallic mold casting, arc melting and unidirectional zone melting, etc.; (5) the high tensile strength, low Young's modulus, and high impact fracture energy of nonferrous metal-based bulk amorphous alloys; (6) the soft magnetic properties of Fe- and Co-based bulk amorphous alloys; (7) hard magnetic properties of Nd- and Pr-based bulk amorphous alloys; (8) the viscous flow and microformability of bulk amorphous alloys in a supercooled liquid region, and (9) future aspects of applications. These new results enable eliminating of the limitation of sample shape which has prevented the development of amorphous alloys as engineering materials. They are expected to give rise to a new era of amorphous alloys.

1.1 History of Bulk Amorphous Alloys

Since an amorphous phase was prepared in the Au-Si system by rapid solidification by Klement et al. in 1960 [1], a great number of scientific and engineering data for amorphous alloys have been accumulated. As a result, it has been clarified that amorphous alloys have new alloy compositions and new atomic configurations which differ from those of crystalline alloys. These features have facilitated the appearance of various characteristics, such as good mechanical properties, useful physical properties and unique chemical properties [2–5] which have not been obtained from conventional crystalline alloys. Recently, amorphous alloys have also attracted increasing interest as precursors to produce nanocrystalline alloys by crystallization because of good mechanical properties [6,7], soft magnetism [8,9], hard magnetism [10–12], high magnetostriction in low applied fields [13], and high catalytic properties [14] which have not been obtained on ordinary amorphous or crystalline alloys. Based on these results, the scientific and engineering importance of amorphous and nanocrystalline alloys has steadily increased for the last three decades. When attention is paid to bulk amorphous alloys with low critical cooling rates (R_c) for glass formation, it is known that $Pd_{40}Ni_{40}P_{20}$ [15] and $Pd_{76}Cu_6Si_{18}$ [16] amorphous alloys can be produced in a bulk form with diameters up to 3 mm and 0.3 mm, respectively, by water quenching. Subsequently, it was noted that a flux treatment using a B_2O_3 medium for the Pd-Ni-P alloy was effective increasing the maximum sample thickness (t_{max}) for glass formation, and the t_{max} reaches about 10 mm [17,18]. Thus, for a long period before 1990, no other bulk amorphous alloys except the Pd- and Pt-metalloid systems were synthesized because of the necessity for R_c to be higher than 10^5 K/s. Recently, a number of bulk amorphous alloys with R_c less than 10^3 K/s have been found in multicomponent alloy systems of Mg-Ln-TM [19], Ln-Al-TM [20], Zr-Al-TM [21,22], Zr-(Ti,Nb,Pd)-Al-TM [23], Zr-Ti-TM-Be [24], Pd-Cu-Ni-P [25], Fe-(Al,Ga)-(P,C,B,Si,Ge) [26,27], Pd-Ni-Fe-P [28], and Fe-(Co,Ni)-(Zr,Nb,Ta)-B [29,30], Fe-(Co,Ni)-(Zr,Nb,Ta)-(Mo,W)-B [31] and Co-Fe-(Zr,Nb,Ta)-B [32] (Ln=lanthanide metal; TM=transition metal). Table 1.1 summarizes the alloy systems of the bulk amorphous alloys, the calendar years when their alloy systems were found, and the t_{max}. Notice that the t_{max} reaches about 30 mm [33] for Zr-Al-TM alloys, 25 mm [34] for Zr-Ti-TM-Be

Table 1.1. Alloy systems, years and references of new multicomponent alloys with high glass-forming ability

I. Nonferrous Metal Base	Years	t_{max}/mm
Mg-Ln-M (Ln=lanthanide metal, M=Ni, Cu or Zn)	1988	10
Ln-Al-TM (TM=VI-VIII transition metal)	1989	10
Ln-Ga-TM	1989	10
Zr-Al-TM	1990	30
Ti-Zr-TM	1993	3
Zr-Ti-TM-Be	1993	25
Zr-Ti-Al-TM	1995	20
Pd-Cu-Ni-P	1996	72
Pd-Cu-B-Si	1997	10
II. Ferrous Group Metal Base	Years	t_{max}/mm
Fe-(Al,Ga)-(P,C,B,Si,Ge)	1995	3
Fe-(Nb,Mo)-(Al,Ga)-(P,B,Si)	1995	3
Co-(Al,Ga)-(P,B,Si)	1996	1
Fe-(Zr,Hf,Nb)-B	1997	5
Co-(Zr,Hf,Nb)-B	1997	1
Ni-(Zr,Hf,Nb)-B	1997	1
Fe-(Co,Ni)-(Zr,Hf,Nb)-B	1997	6

alloys, 72 mm [35] for the Pd-Cu-Ni-P alloy, 25 mm [28] for the Pd-Ni-Fe-P alloy, and 6 mm [31] for the Fe-Co-Zr-Mo-W-B alloy. This paper is intended to review our recent results on the fabrication, fundamental properties, workability, and applications of the new bulk amorphous alloys with thicknesses above several millimeters which were mainly obtained by our group during the last eight years.

1.2 Dominant Factors for High Glass-Forming Ability

To an amorphous phase by rapid solidification is essential to suppress the nucleation and growth reaction of a crystalline phase in the supercooled liquid region between the melting temperature (T_m) and the glass transition temperature (T_g). The R_c has been reported [36] to be higher than 10^4 K/s for Fe-, Co- and Ni-based amorphous alloys and higher than 10^2 K/s for Pd- and Pt-based amorphous alloys, much higher than those for oxide glasses.

Recently, we have succeeded [37–41] in finding new multicomponent amorphous alloys with much lower R_c ranging from 0.100 K/s to several hundreds K/s, as shown in Fig. 1.1a. Simultaneously, the t_{max} increases drastically from several millimeters to about one hundred millimeters. Note that the lowest R_c and the largest t_{max} are almost comparable to those for oxide and fluoride glasses. Figure 1.1a also shows that the new amorphous alloys have a higher reduced glass transition temperature (T_g/T_m) above 0.60. Furthermore, the new amorphous alloys have a much wider supercooled liquid region which is defined by the difference between crystallization temperature (T_x) and T_g, $\Delta T_x (=T_x-T_g)$, as shown in Fig. 1.1b. From the empirical relationships between R_c, t_{max}, and T_g/T_m or ΔT_x, high glass-forming ability (GFA) is obtained by satisfying the following two factors; (1) high T_g/T_m, and (2) large ΔT_x. As summarized in Table 1.1, the bulk amorphous alloys consist of multicomponent systems containing Mg, Ln, Zr, Pd, Fe, or Co as a main constituent element. Based on the multicomponents of the amorphous alloys with high GFA, we have proposed [37–41] three simple empirical rules for achieving high GFA for metallic alloys as follows: (1) multicomponent system consisting of more than three elements, (2) significant difference in atomic size ratios above about 12% among the main constituent elements, and (3) negative heats of mixing among their elements. The importance of the three empirical rules can be interpreted from a number of experimental data as well as theory. Various models to describe glass formation and GFA have been proposed to date [36]. These models can be grouped into three categories of thermodynamics, kinetics and structure, depending on the factors that are viewed as decisive in forming amorphous alloys. Therefore, the reason that the above-described ternary amorphous alloys have lower R_c must be consistently understood on the basis of these three factors. First, we investigate the reason for the high GFA from the thermodynamic point of view. It is known that a high GFA is obtained in the condition of low free energy

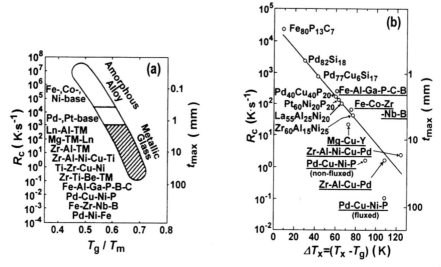

Fig. 1.1 a,b. Relationship between the minimum critical cooling rate for glass formation (R_c), the maximum sample thickness (t_{max}), and the reduced glass transition temperature (T_g/T_m) or the temperature interval of supercooled liquid region $\Delta T_x (= T_x - T_g)$ for new multicomponent amorphous alloys in lanthanide(Ln)-, Mg-, Zr-, Pd-Cu-, Fe-, Co- and Ni-based systems. The data for previously reported metallic amorphous alloys are also shown for comparison

$\Delta G(T)$ for transforming liquid to the crystalline phase. In the relationships $\Delta G = \Delta H_f - T \Delta S_f$, for Gibbs free energy, a low ΔG value is obtained in cases of low ΔH_f and large ΔS_f. Here, the ΔH_f and ΔS_f are enthalpy of fusion and entropy of fusion, respectively. A large ΔS_f is expected to be obtained in multicomponent systems because ΔS_f is proportional to the number of microscopic states. The free energy at a constant temperature also decreases in cases of low chemical potential caused by low enthalpy and high T_g/T_m as well as high liquid/solid interfacial energy. Based on these thermodynamic factors, it is concluded that the multiplication of alloy components leading to the increase of ΔS_f increases the degree of dense random packing which is favorable for decreasing ΔH_f and increasing of solid/liquid interfacial energy. This interpretation is consistent with the present result that a high GFA has been obtained for the above-described multicomponent systems containing more than three elements. Furthermore, it has been recently made clear that the multicomponent amorphous alloys have high packing fractions, as evidenced from the low specific volumes shown in 7. Secondly, the reason for the high GFA is investigated from the kinetic point of view. The homogeneous nucleation and growth of a crystalline phase with a spherical morphology from a supercooled liquid are expressed by the following relationships [42]:

$$I = \frac{10^{30}}{\eta}\exp[-b\alpha^3\beta/(T_r(1-T_r)^2)] \quad [\text{cm}^{-3}\text{s}^{-1}] \tag{1.1}$$

$$U = \frac{10^2 f}{\eta}[1 - \exp(-\beta\Delta T_r/T_r(T/T_m))] \quad [\text{cms}^{-1}]. \tag{1.2}$$

Here, T_r is the reduced temperature (T/T_m), ΔT_r is the difference in temperature from T_m, b is a shape factor and is $16\pi/3$ for a spherical nucleus, η is viscosity, and f is the fraction of nucleus sites at the growth interface. α and β are dimensionless parameters related to the liquid/solid interfacial energy (σ), while ΔH_f and ΔS_f can be expressed as $\alpha = (NoV)^{1/3}\sigma/\Delta H_f$ and $\beta = \Delta S_f/R$. Here, No, V, and R are the Avogadro number, the atomic volume, and the gas constant, respectively. In these relationships, the important parameters are η, α, and β. The increase in the three parameters decreases the I and U values, leading to an increase of GFA. The increases of α and β also imply an increase in σ and ΔS_f and a decrease in ΔH_f, consistent with the interpretation of achieving a high GFA derived from the thermodynamic point of view. Furthermore, one can notice that h is closely related to the T_g/T_m value and $\alpha^3\beta$ reflects the thermal stability of the supercooled liquid. The importance of the $\alpha^3\beta$ parameter can be understood from the following two examples; (1) when the value of $\alpha\beta^{1/3}$ exceeds 0.9, unseeded liquid would not in practice crystallize at any cooling rate, and (2) when $\alpha\beta^{1/3}$ is below 0.3, the liquid would be impossible to suppress crystallization. Furthermore, the reason for the high GFA of the particular multicomponent alloys is investigated from the structural point of view. As described above, the present multicomponent amorphous alloys consist of the elements with significant difference in atomic sizes above about 12% and negative heats of mixing, that is, the atomic sizes of the constituent elements in the ternary amorphous alloys can be classified into the three groups of larger, intermediate, and smaller, as shown in Fig. 1.2. The combination of the significant difference in the atomic sizes and negative heats of mixing is expected to induce a denser random packing fraction in the supercooled liquid which enables achieving high liquid/solid interfacial energy as well as the difficulty of atomic rearrangements leading to low atomic diffusivity and high viscosity. The formation of a denser random packed structure has also been confirmed from the X-ray diffraction profiles of the ternary La-Al-Ni [43], Mg-Ni-La [44], and Zr-Al-Ni [45] amorphous alloys, that is, no pre-peak at the lower angle side or any separation on the second peak is seen in the X-ray interference functions [43]. The feature of the X-ray interference functions is analogous to that for a liquid phase, indicating that the ternary amorphous alloy has a homogeneously mixed atomic configuration corresponding to the formation of a denser random packed structure. The increase in the random packing density is also supported by the very

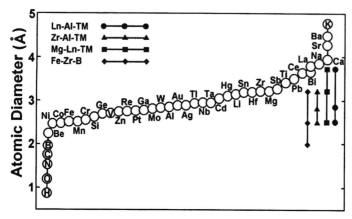

Fig. 1.2. Atomic radii of the constituent elements for the new ternary amorphous alloys

small change in density of the multicomponent amorphous alloys upon crystallization [46]. The density of the $Zr_{60}Al_{10}Ni_{10}Cu_{20}$ and $Pd_{40}Cu_{30}Ni_{10}P_{20}$ alloys is 6.72 and 9.27 Mg/m^3, respectively, for the amorphous state and 6.73 and 9.31 Mg/m^3, respectively, for the crystalline state. The change in density upon crystallization is 0.30 and 0.54%, respectively, which are much smaller than those (about 2%) [2,3,47] for ordinary amorphous alloys.

We have also evaluated the changes in the atomic distances and coordination numbers among the constituent elements upon crystallization of the ternary amorphous alloys from the radial distribution function(RDF) data obtained by the anomalous X-ray scattering technique [43–45]. Table 1.2 summarizes the atomic distances and coordination numbers of each atomic pair for an amorphous $Zr_{60}Al_{15}Ni_{25}$ alloy in as-quenched and crystallized states [45]. Although no significant changes in the coordination numbers and atomic distances of Ni-Zr and Zr-Zr pairs upon crystallization are recognized, one can notice the drastic change in the coordination number of the Zr-Al atomic pair upon crystallization. The significant change indicates that the local atomic configurations in the amorphous alloy are significantly different from those in the corresponding equilibrium crystalline phase. This result indicates the necessity of long-range rearrangement of Al around Zr in the course of crystallization. The long-range atomic rearrangement is difficult in the denser random packed amorphous phase. Based on the above-described results and discussion, Fig. 1.3 summarizes the reasons for achieving a high GFA of the multicomponent amorphous alloys in Ln-, Mg-, Zr-, Pd-, Fe-, and Co-based systems [37–41]. It is again confirmed that the high GFA is attributable to the formation of a new kind of supercooled liquid with a random packing density, new short-range atomic configurations, and long-range atomic interactions resulting from multicomponent elements with different atomic sizes and negative heats of mixing. The peculiar liquid structure can

Table 1.2. Atomic distances (r) and coordination numbers (N) calculated from (a) the ordinary radial distribution function (RDF) and from the environmental RDFs for (b) Zr and (c) Ni in the as-quenched and annealed $Zr_{60}Al_{15}Ni_{25}$ amorphous alloys

Alloy		r_1/nm	N_{ZrNi}	r_2/nm	N_{ZrZr}	N_{ZrAl}
As-Quenched	(a)	0.267±0.002	2.3±0.2	0.317±0.002	10.3±0.7	−0.1±0.9
	(b)	0.267±0.002	2.1±0.2	-	-	-
	(c)	0.269±0.002	2.3±0.2	-	-	-
Crystallized	(a)	0.268±0.002	3.0±0.2	0.322±0.002	8.2±0.7	0.8±0.9
	(b)	0.267±0.002	3.0±0.2	-	-	-
	(c)	0.273±0.002	2.3±0.2	-	-	-

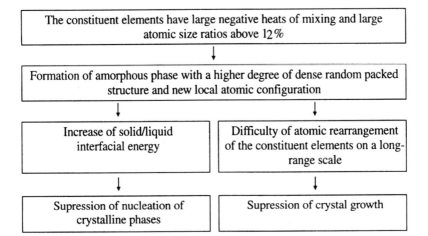

Fig. 1.3. Summary of the reasons for achieving a high glass-forming ability for some ternary alloy systems, such as La-Al-TM, Mg-Ln-TM, and Zr-Al-TM (Ln=lanthanide metal, TM=transition metal)

have a higher liquid/solid interfacial energy leading to the suppression of nucleation of a crystalline phase, a higher viscosity leading to an increase in T_g/T_m, and a lower atomic diffusivity leading to the difficulty of growing a crystalline phase.

1.3 Crystal Nucleation and Growth Behavior of Alloys with High GFA

It has been pointed out in Sect. 1.2 that the Al element plays an important role in increasing the GFA of Zr-Al-TM amorphous alloys. We have further examined the reason for the effect of Al addition on the stability of the supercooled liquid to crystallization. In the compositional dependence of T_g and T_x for the amorphous $Zr_{65}Al_xCu_{35-x}$ alloys [48], T_g increases linearly with increasing Al content, and T_x shows a maximum at 7.5 at % Al. Thus, the extension of the supercooled liquid region is dominated by the retardation of crystallization. Clarification of the crystallization behavior from the highly supercooled liquid is important for understanding the origin of the high thermal stability of the supercooled liquid. Inoue et al. have confirmed [48] that crystallization is completed by the simultaneous precipitation of Zr_2Ni, Zr_2Cu, $ZrNi$ and $ZrAl$ phases and the kinds of phases are independent of annealing temperature (T_a). Precipitation of the crystalline phases during isothermal annealing takes place through a single exothermic reaction. The incubation time (τ) for precipitation of their crystalline phases was also measured at various T_a for $Zr_{67}Cu_{33}$, $Zr_{65}Al_{7.5}Cu_{27.5}$, and $Zr_{65}Al_{20}Cu_{15}$ amorphous alloys. A positive deviation of τ from the linear relationship in the supercooled liquid region was seen only for the $Zr_{65}Al_{7.5}Cu_{27.5}$ alloy with the widest ΔT_x, indicating the suppression of the precipitation of their crystalline phases. Similar positive deviation in the supercooled liquid has also been recognized [48] for η which was determined from the relationship of $\tau \propto T\,\eta/(T_m-T)^2 f$ [49] for the amorphous $Zr_{65}Al_{7.5}Cu_{27.5}$ alloy. Therefore, the supercooled liquid of the Zr-Al-Cu alloy has a unique atomic configuration which increases h and decreases atomic diffusivity, in agreement with the interpretation described in Sect. 1.2.

Inoue et al. also determined [39] the peak temperatures of the nucleation and growth reactions of the precipitation phases for the amorphous $Zr_{67}Cu_{33}$, $Zr_{65}Al_{7.5}Cu_{27.5}$, and $Zr_{65}Al_{7.5}Ni_{10}Cu_{17.5}$ alloys by measuring the change in the peak temperature of the exothermic reaction due to crystallization as a function of T_a for the samples subjected to preannealing for 60 s at various T_a. Figure 1.4 plots the difference in the inverse of the peak temperature between the preannealed and the as-quenched samples ($1/T_p - 1/T_p^o$) as a function of T_a. A maximum occurs for the $Zr_{65}Al_{7.5}Cu_{27.5}$ and $Zr_{65}Al_{7.5}Ni_{10}Cu_{17.5}$ alloys with higher GFA, but no maximum is seen for the $Zr_{70}Cu_{30}$ alloy. The maximum temperature, previously reported [50], corresponds to the peak temperature of the nucleation reaction of the crystalline phase. The T_x shown in Fig. 1.5 is measured at the fastest heating rate of 5.33 K/s in the DSC analysis and can be regarded as the peak temperature of the growth phase of the precipitates. Thus, the Zr-Al-Cu and Zr-Al-Ni-Cu alloys have a significant difference in peak temperatures between the nucleation and growth phases, and the difference becomes more significant with the multiplication of alloy components. It is known that the distinct difference in the peak temperatures

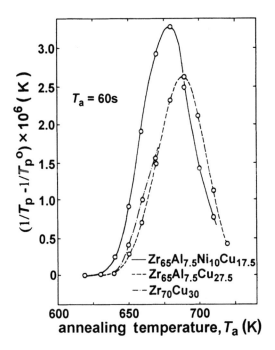

Fig. 1.4. Relationship between $(1/T_p - 1/T_p^o)$ and T_a for amorphous $Zr_{70}Cu_{30}$, $Zr_{65}Al_{7.5}Cu_{27.5}$, and $Zr_{65}Al_{7.5}Ni_{10}Cu_{17.5}$ alloys

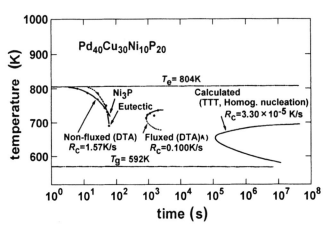

Fig. 1.5. Continuous cooling transformation (C.C.T.) curve of the $Pd_{40}Cu_{30}Ni_{10}P_{20}$ alloy subjected to the B_2O_3 flux treatment, compared with that for the same alloy without the flux treatment. The temperature-time-transformation (T.T.T.) curve calculated on the basis of the homogeneous nucleation and growth theories is also presented for comparison

of the nucleation and growth phases increases the GFA by an increase in the thermal stability of the supercooled liquid, consistent with the present result. By utilizing the difference in the peak temperatures of the nucleation and growth phases, tailored control of the particle size of the crystallization-induced phase has been reported [48] for the $Zr_{65}Al_{7.5}Cu_{27.5}$ alloy.

1.4 Continuous Cooling Transformation of Alloys with High GFA

It is important to know R_c to determine GFA and to produce a bulk amorphous alloy. The use of the multicomponent alloys has enabled the actual measurement of a continuous cooling transformation (C.C.T.) curve by using various kinds of thermocouples, pyrometers, and DTAs. In particular, a new Pd-Cu-Ni-P bulk amorphous alloy was produced by slow cooling in the DTA where the cooling behavior of the melted ingot is easily measured in a wide cooling rate range and the precipitation of a crystalline phase can be detected on the DTA curve. Figure 1.5 shows the C.C.T. curve of the $Pd_{40}Cu_{30}Ni_{10}P_{20}$ alloy obtained at various cooling rates in the DTA [51]. The amount of the ingot was 2 g and the cooling rate was changed in the range from 0.033 to 3.74 K/s. Assuming that R_c is a minimum cooling rate to avoid the intersection with the C.C.T. curve for the liquid, R_c of the Pd-Cu-Ni-P alloy is measured as 1.58 K/s [51] for the nonfluxed melt and 0.100 K/s [52] for the B_2O_3 fluxed melt. The flux treatment is effective for decreasing R_c because the molten alloy is cleaned by the flux treatment and heterogeneous nucleation sites are reduced [16–18]. The R_c value is believed to be the lowest for all metallic amorphous alloys reported to date.

The effect of the flux treatment on the suppression of heterogeneous nucleation has been determined by comparing the present experimental C.C.T. curve with the predicted time-temperature-transformation (T.T.T.) curve which can be calculated from homogeneous nucleation and growth theories [53,54]. When the homogeneous nucleation and growth rates of crystal are represented by I_{vhomo} and U_c, respectively, the volume fraction (X) crystallized in time (t) is expressed by (1.3) in the Johnson-Mehl-Avrami treatment of transformation kinetics:

$$X = \left(\frac{\pi}{3}\right) I_{vhomo} U_c^3 t^4 \tag{1.3}$$

I_{vhomo} and U_c for glass-forming systems have been given by (1.1) and (1.2) [55], respectively. The relationship between D and η is expressed by the following Stokes-Einstein equation

$$D = kT/3\pi a_0 \eta . \tag{1.4}$$

By inserting (1.1), (1.2) and (1.4) in (1.3), (1.5) can be obtained.

$$t = \frac{9.3\eta}{kT}\left[\frac{Xa_0^9}{f^3 N_v}\exp(1.07/T_r^3\Delta T_r^2)/\{1-\exp(-\Delta H_m^f \Delta T_r/RT)\}^3\right]^{1/4} . \tag{1.5}$$

Calculated t was from the following experimental [56] and reference [53,57] values: $f=0.2T_r$, $N_v=8.01\times10^2$ atom/cm^3, $\Delta H_m^f=4840$J/mol, $a_o=0.133$nm, $x=10^{-6}$, and $\eta =9.34\times10^{-3}\exp[4135/(T-447)]$. Here, N_v and a_o are the calculated values from the atomic sizes and atomic percentages of the constituent elements for Pd$_{40}$Cu$_{30}$Ni$_{10}$P$_{20}$. The ΔH_m^f and h are the experimental data for the Pd$_{40}$Cu$_{30}$Ni$_{10}$P$_{20}$ amorphous alloy. The relationship between t and T obtained thus is also shown in Fig. 1.5, where the experimental data (C.C.T. curve) of the Pd$_{40}$Cu$_{30}$Ni$_{10}$P$_{20}$ alloy are also shown for comparison. The calculated T.T.T. curve lies at the longer time and lower temperature side compared with the experimental C.C.T. curve. The nose point on the calculated T.T.T. curve is located at $T=670$ K and $t=10^5$ s, and the R_c is determined to be 1.34×10^{-3} K/s which is smaller by two orders than that ($R_c=0.100$ K/s) for the fluxed Pd$_{40}$Cu$_{30}$Ni$_{10}$P$_{20}$ alloy. The significant difference in R_c implies that the B$_2$O$_3$ flux treatment cannot completely eliminate the heterogeneous nucleation sites in the Pd-based alloy liquid. If we can completely eliminate the heterogeneous nucleation sites, it is predicted that the $t_{\max}$ of the resulting bulk amorphous alloy will reach as much as about 1.5 m [52] even for the Pd$_{40}$Cu$_{30}$Ni$_{10}$P$_{20}$ alloy. Similar C.C.T. curves are obtained for Zr-Al-Ni-Cu amorphous alloys. R_c values have been reported to be about 10 K/s for the arc-melted ingot [58], 40 K/s for the unidirectionally solidified ingot [59], and 80 K/s for the ingot cast into a wedge-shaped copper mold [60]. The difference in R_c among the solidification processes reflects the difference in the ease of heterogeneous nucleation of a crystalline phase.

1.5 Preparative Methods and Maximum Thickness of Bulk Amorphous Alloys

By choosing appropriate compositions with large ΔT_x, high T_g/T_m, and low R_c, bulk amorphous alloys have been produced by various techniques which are classified into two groups, solidification and consolidation. For the solidification technique, one can list water quenching, high pressure die casting, arc melting, copper mold casting, unidirectional casting, and suction casting. Bulk amorphous alloys are also produced by hot pressing and warm extrusion of atomized amorphous powders in the supercooled liquid state [37–41]. Table 1.3 summarizes the typical alloy systems, production methods, maximum sample thicknesses, and approximate critical cooling rates. Here, it is important for understanding of high thermal stability of the supercooled liquid to examine the ease of movement of the amorphous/crystal interface in melted ingots. The quaternary and penternary alloys in the Zr-Al-TM (TM=Co, Ni, Cu) systems can be made into a button-shape bulk amorphous form by arc melting on a copper hearth [61]. However, in this melting method, it is difficult to suppress completely the precipitation of crystalline phases because of the ease of heterogeneous nucleation resulting from incomplete melting at the bottom in contact with the copper hearth. Here, it is

Table 1.3. Alloy systems, production methods, and maximum sample thickness (t_{max}) of bulk amorphous alloys

Production Methods of Bulk Amorphous Alloys	
I. Solidification	II. Consolidation
1. Water quenching	1. Hot pressing
2. Copper mold casting	2. Warm extrusion
3. High-pressure die casting	
4. Arc melting	
5. Undirectional casting	
6. Suction casting	

Maximum Thickness (t_{max}) at Cooling Rate (R_c) of Bulk Amorphous Alloys	
Alloy system	t_{max} (mm)
Ln-Al-(Cu,Ni)	$\cong 10$
Mg-Ln-(Cu,Ni)	$\cong 10$
Zr-Al-(Cu,Ni)	$\cong 30$
Zr-Ti-Al-(Cu,Ni)	$\cong 30$
Zr-Ti-(Cu-Ni)-Be	$\cong 30$
Fe-(Al,Ga)-(P,C,B,Si)	$\cong 3$
Pd-Cu-Ni-P	$\cong 75$
Fe-(Co,Ni)-(Zr,Hf,Nb)-B	$\cong 6$

noted that only an amorphous phase is formed in the inner region where the heat flux reinforced from the copper hearth disappears and the cooling capacity is reduced, that is, the low cooling rate obtained by arc melting is high enough to form an amorphous phase in the Zr-Al-Co-Ni-Cu system. The crystalline phase can be divided into three different zones, on the outer equiaxed zone, the columnar dendrite zone, and the metastable single phase zone, as illustrated in Fig. 1.6. It has further been confirmed that the interface between the amorphous and crystalline phases is very smooth because the highly supercooled solidification condition defined by the criterion $V_i \gg D_i/\delta_i$ [62] is satisfied. Here, V_i is the velocity of the liquid/solid interface, D_i is the diffusivity of the constituent elements at the interface and d_i is the distance of the atomic rearrangement required for formation the solid phase from the liquid. On the other hand, in the inner region which is far from the copper hearth, one cannot take account of the effect of reinforced heat flux from the copper hearth. Therefore, it is expected [63] that the direction of heat flux agrees with that of grain growth, and the interface becomes

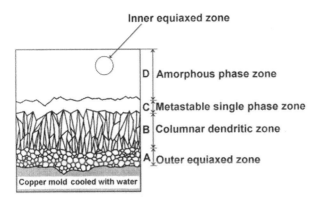

Fig. 1.6. Schematic illustration showing the features and dominant parameters of the solidification structure of an arc-melted ingot

unstable and irregular. However, we can always observe a smooth interface even in the inner region [61]. Considering that the wave length (λ_i) of the interface can be expressed by $\lambda_i = 2\pi(D\Gamma/V\Delta T_o)^{1/2}$, ($\Gamma = \sigma/\Delta S_f$, and ΔT_o is the temperature interval between liquidus and solidus temperatures) [64], the inconsistency is presumably because the present supercooled liquid has a high liquid/solid interfacial energy. This presumption is consistent with the interpretation (Sect. 1.2) derived from the thermodynamic factors. Zr-based bulk amorphous alloys have been expected to develop as an important engineering material. Therefore it is important to search for an additional element which increases the GFA by suppressing of heterogeneous nucleation and growth phases of the crystalline phase. The T_g, T_x, and ΔT_x of the melt-spun $Zr_{65-x}Al_{10}Cu_{15}Ni_{10}M_x$ (M=Ti or Hf) amorphous alloys were examined as a function of M content [23]. No distinct changes in T_g, T_x, and ΔT_x are seen for Hf, but the addition of Ti significantly decreases T_x and slightly increases T_g, leading to a decrease of ΔT_x from 107 K at 0% Ti to 38 K at 10% Ti. The influence on ΔT_x is significantly different between Ti and Hf. Similarly, the T_g, T_x, and ΔT_x values as a function of additional M elements were examined for $(Zr_{0.65}A_{0.1}Cu_{0.15}Ni_{0.1})_{100-x}M_x$ (M=V, Nb, Cr, or Mo), $Zr_{65}Al_{10}Cu_{15-x}Ni_{10}M_x$ (M=Fe, Co, Pd, or Ag) and $Zr_{65}Al_{10}Cu_{15}Ni_{10-x}M_x$ (M=Fe, Co, Pd, or Ag) alloys. As these additional elements increase, T_x decreases gradually, and T_g increases slightly, leading to a significant decrease in ΔT_x. This tendency is independent of the kind of the M element. No element was effective in increasing ΔT_x. Because the increase in T_g by the addition of the M elements is slight, we concluded that the significant decrease of ΔT_x for the M-containing alloys is due to the decrease in T_x. Subsequently, the formative tendency of an amorphous phase in arc-melted $Zr_{65}Al_{10}Cu_{15}Ni_{10}$ alloy ingots with a t_{max} of about 8 mm was examined in comparison with the magnitude of ΔT_x for melt-spun amorphous ribbons. Figure 1.7 shows the change in the ratio of amorphous area to the total cross-sectional area with

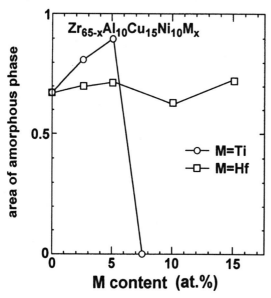

Fig. 1.7. Changes in the area ratio of amorphous phase with M content for arc-melted $Zr_{65-x}Al_{10}Cu_{15}Ni_{10}M_x$ (M=Ti or Hf) ingots with a maximum thickness of 8 mm

Ti or Hf content for arc-melted $Zr_{65-x}Al_{10}Cu_{15}Ni_{10}M_x$ (M=Ti or Hf) ingots. The area ratio of the amorphous phase remains almost constant ($\equiv 70\%$) for M=Hf. However, the ratio for the Ti-containing alloys increases from 67% for the quaternary alloy without M elements to about 90% for the 5% Ti alloy and then decreases with further increasing Ti content. Note that the area ratio of the amorphous phase increases significantly for the Ti-containing alloys with smaller ΔT_x values. This striking result suggests that the formative tendency of an amorphous phase in an arc-melted ingot is not directly related to ΔT_x. The change in the area ratio of the amorphous phase in arc-melted ingots by the addition of other M elements has also been examined. A similar increase in the area ratio of the amorphous phase to about 90% occurs for Nb in the $(Zr_{0.65}Al_{0.1}Cu_{0.15}Ni_{0.1})_{100-x}M_x$ (M=V, Nb, Cr, or Mo) alloys and Pd in the $Zr_{65}Al_{10}Cu_{15-x}Ni_{10}M_x$ and $Zr_{65}Al_{10}Cu_{15}Ni_{10-x}M_x$ (M=Fe, Co, Pd, or Ag) alloys, though other elements except Nb and Pd significantly decrease the area ratio of the amorphous phase. We also examined optical micrographs of the cross-sectional structure in the Ti-, Nb-, and Pd-containing alloys with high area ratios of the amorphous phase, together with the data of the Zr-Al-Cu-Ni quaternary alloy. It was clearly seen that the area ratio of amorphous phase in the Zr-based alloys increases drastically by the addition of 5% Ti, 2.5% Nb, or 5% Pd and the formation of the crystalline phase is limited to the small region in contact with the copper hearth. On the other hand, optical micrographs of the $Zr_{65}Al_{10}Cu_{15}Ni_{10-x}Co_x$ ($x=0, 2.5, 5, 7.5,$ and 10 at %) alloys revealed a drastic decrease in the area ratio of the amorphous phase with increasing Co content, and no amorphous region was seen for the 10% Co alloy.

We also examined the relationship between the composition range in which an amorphous phase is formed in the arc-melted ingots and the temperature interval of the supercooled liquid region (ΔT_x). The amorphous phase was formed even at the compositions with smaller ΔT_x values only for the Ti-, Nb-, and Pd-containing alloys. The minimum ΔT_x value for the formation of the amorphous phase in arc-melted ingots is 54 K for the Ti-containing alloy, 65 K for the Nb-containing alloy, and 47 to 49 K for the Pd-containing alloy, indicating that the effectiveness of the additional elements is the greatest for Pd, followed by Ti, and then Nb. Again note that the $Zr_{65}Al_{10}Cu_{15-x}Ni_{10}Pd_x$ ($x \leq 7.5$ at %) and $Zr_{65}Al_{10}Cu_{15}Ni_{10-x}Pd_x$ ($x \leq 5$ at %) alloys with ΔT_x less than 50 K are made in bulk amorphous form with a t_{max} of about 8 mm by arc melting. These results indicate clearly that the formative tendency of an amorphous phase by arc melting is not directly related to ΔT_x values, and differs from the previous tendency [43–49] for R_c to decrease with increasing ΔT_x. The reason for the significant glass formation of the Zr-based alloy ingots with small ΔT_x values has been investigated [25]. The inconsistency is thought to result from incomplete melting of the arc-melted ingot in the region in contact with the copper mold. The temperature in the region in contact with the copper mold seems to lie around T_m because the temperature of the copper mold is usually below the T_m of Cu metal. The incomplete melting at the contact site induces the residual existence of crystalline phases which act as crystal nucleation sites in the subsequent solidification process. The arc-melted ingot always contains a number of crystalline nuclei in the region in contact with the copper mold. Therefore, it is thought that the GFA of the arc-melted ingot is not dominated by the difficulty of generating of a crystalline nucleus, but is attributable to the difficulty of the growth phase of a crystalline nucleus. The determination of the GFA by ΔT_x simultaneously contains both parameters of nucleation and growth phases. On the other hand, determination of glass formation from the area ratio of the amorphous phase in arc-melted ingots contains only the parameter of the growth phases. The addition of Ti, Nb, or Pd into the Zr-Al-Cu-Ni alloy is useful for suppressing of the growth phase of crystalline nuclei.

1.6 Structural Relaxation and Glass Transition

Figure 1.8 shows the change in the thermograms of a $Zr_{65}Al_{7.5}Cu_{27.5}$ amorphous alloy with T_a during an annealing period t_a of 12 h [65]. On heating the as-quenched sample, the apparent specific heat ($C_{p,q}$) begins to decrease, indicative of a structural relaxation at 380 K, and shows a minimum value at 575 K. With a further increase in temperature, $C_{p,q}$, increases gradually up to 620 K and then increases rapidly in the glass transition range from 630 to 675 K, reaching 47.4 J/mol·K for the supercooled liquid at around

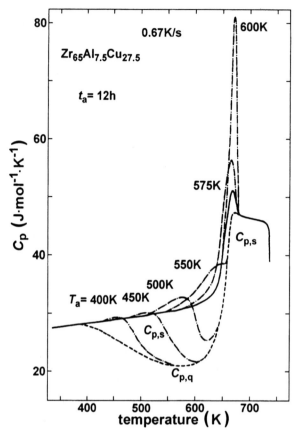

Fig. 1.8. The thermograms for an amorphous $Zr_{65}Al_{7.5}Cu_{27.5}$ alloy annealed for 12 h at temperatures from 400 to 620 K. The solid line represents the thermogram of the sample heated to 690 K.

680 K. Then it decreases rapidly due to crystallization at 731 K. The heating curve for the annealed sample shows a $C_{p,a}(T)$ behavior that closely follows the C_p curve of the reference sample, $C_{p,s}$, up to each T_a and then exhibits an excess endothermic relative to the reference sample before merging with that of the as-quenched sample in the supercooled liquid region above 642 K. The main features of Fig. 1.8 are summarized as follows. (1) The sample annealed at T_g shows an excess endothermic specific heat beginning at T_a, implying that the C_p value in the temperature range above T_a depends on the thermal history and consists of configurational contributions, as well as those arising from purely thermal vibrations. (2) The magnitude of the endothermic peak increases rapidly at T_a just below T_g. (3) The excess endothermic peak is recoverable, but the exothermic broad peak is irrecoverable, and the $C_{p,a}(T)$ curves for the annealed samples couple the recoverable endothermic

and irrecoverable exothermic reactions. The excess endothermic peak has been thought [66] to occur by rearrangement from a relaxed atomic configuration caused by annealing to an atomic configuration that is more stable at temperatures above T_a. The excess endothermic reaction reflects the atomic configuration caused by annealing, and hence we can obtain information on the structural relaxation during annealing by examining the change in the excess endothermic peak with T_a and t_a. The temperature dependence of the differences in C_p between the annealed and the reference states, [$\Delta C_{p,\text{endo}} = C_{p,a}(T) - C_{p,s}(T)$], for $\text{La}_{55}\text{Al}_{25}\text{Ni}_{20}$ and $\text{Zr}_{65}\text{Al}_{7.5}\text{Cu}_{27.5}$ alloys is shown in Fig. 1.9. With increasing T_a, $\Delta C_{p,\text{max}}$ for the two alloys initially increases gradually, followed by a rapid increase at temperatures slightly below T_g. The rapid increase in $\Delta C_{p,\text{max}}$ is interpreted as corresponding to glass transition phenomenon. Similarly, the rapid decrease in $\Delta C_{p,\text{max}}$ above T_g results from achieving of internal equilibrium due to the very short relaxation times in the supercooled liquid region. Although the change in $\Delta C_{p,\text{max}}$ with T_a is similar to that for Zr-Cu and Zr-Ni amorphous alloys [67], it is different from the two-stage change which has been observed [68] for all amorphous alloys in metal-metalloid systems containing more than two types of metallic elements. It has been thought [68] that the appearance of the two-stage relaxation process is due to the difference in the relaxation time between the

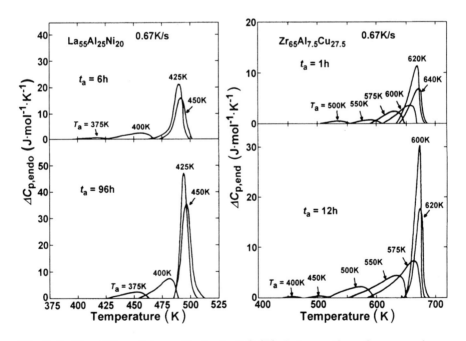

Fig. 1.9. The differential specific heat, $\Delta C_p(T)$, between the reference and annealed samples, for amorphous $\text{La}_{55}\text{Al}_{25}\text{Ni}_{20}$ and $\text{Zr}_{65}\text{Al}_{7.5}\text{Cu}_{27.5}$ alloys annealed for different periods at temperatures ranging from 375 to 620 K

metal-metal pairs with weaker bonding and the metal-metalloid pairs with stronger bonding. Because no splitting of $\Delta C_{p,\mathrm{max}}$ into two stages is seen as a function of T_a for the La-Al-Ni and Zr-Al-Cu amorphous alloys, the relaxation times seem to be nearly the same for the La-Al, Al-Ni, and La-Ni pairs and for the Zr-Al, Al-Cu, and Zr-Cu pairs, which have large negative enthalpies of mixing. This result suggests that there is no appreciable difference in the attractive bonding between atomic pairs and that the constituent atoms in these amorphous alloys are in an optimum bonding state. The optimum bonding state prevents easy atomic movement, even in the supercooled liquid, and suppresses the nucleation and growth of a crystalline phase, leading to the appearance of the extremely wide supercooled liquid region. The glass transition behavior was examined from the change in $C_p(T)$ by the transition of an amorphous solid to a supercooled liquid and the temperature dependence of the specific heat in the amorphous solid and supercooled liquid. As an example, Fig. 1.10 shows the thermograms of an amorphous $Zr_{60}Al_{15}Ni_{25}$ alloy [21]. The C_p value increases gradually and begins to decrease, indicating an irreversible structural relaxation at 440 K. With a further increase in temperature, the C_p value reaches its minimum at 640 K, then increases rapidly in the glass transition range from 660 to 720 K, and reaches 33.7 J/mol·K for the supercooled liquid around 730 K. With further increased temperature, the C_p value of the supercooled liquid decreases gradually and then rapidly

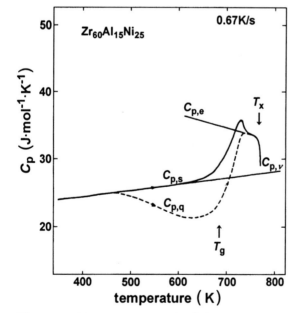

Fig. 1.10. The thermogram $C_{p,q}(T)$ of an amorphous $Zr_{60}Al_{15}Ni_{25}$ alloy in the as-quenched state. The solid line represents the thermogram $C_{p,s}(T)$ of the sample heated to 750 K

due to crystallization at 770 K. It is seen in Fig. 1.10 that the transition of the amorphous solid to the supercooled liquid occurs accompanied by a large increase of $\triangle C_{p,s \to l}$ of 6.25 J/mol·K. The difference in $C_p(T)$ between the as-quenched and the reheated states, $[\triangle C_p(T)]$, manifests the irreversible structural relaxation which is presumed to arise from the annihilation of various kinds of quenched-in "defects" and the enhancement of the topological and chemical short-range ordering by the atomic rearrangement. The $C_{p,s}$ curve of the reheated (control) sample is unaffected by thermal changes and consists of configurational contributions, as well as those arising from purely thermal vibrations. Therefore, the vibrational specific heat, C_p, for the amorphous alloy is extrapolated from C_p values in the low temperature region and is a linear function of temperature:

$$C_p = 23.9 + 1.29 \times 10^{-2}(T - 340) \quad 340 \leq T \leq 620 \,. \tag{1.6}$$

Similarly, the equilibrium specific heat, $C_{p,s}$, of the supercooled liquid, including the vibrational and configurational specific heat, can be expressed by (1.7) based on the data shown in Fig. 1.10,

$$C_{p,e} = 33.7 + 2.59 \times 10^{-2}(755 - T) \quad 735 \leq T \leq 760. \tag{1.7}$$

The $\triangle C_{p,s \to l}$ value for the Zr-Al-Ni amorphous alloys was examined as a function of composition. It was found that the $\triangle C_{p,s \to l}$ value is about 6.5 J/mol in the vicinity of $Zr_{60}Al_{20}Ni_{20}$ and tends to decrease with a deviation from the alloy composition, similar to the compositional dependence of $\triangle T_x$ and $\triangle H_x$, that is, there is a tendency for $\triangle C_{p,s \to l}$ to increase with increasing $\triangle T_x$ and $\triangle H_x$. However, the $\triangle C_{p,s \to l}$ values are considerably smaller than those (10 to 20 J/mol·K) for Pt-Ni-P [69], Pd-Ni-P [69], Mg-Ni-La [20], and La-Al-Ni [21] amorphous alloys. Although the reason for the small $\triangle C_{p,s \to l}$ value remains unknown, it may be related to the result that the glass transition appears at temperatures much higher than the T_g values of the Pt-, Pd-, Mg-, and La-based amorphous alloys. Figure 1.10 also shows the thermograms of the $Zr_{65}Al_{7.5}Ni_{10}Cu_{17.5}$ amorphous alloy with the largest $\triangle T_x$ value of 127 K [70]. The C_p value of the as-quenched phase is 26.9 J/mol·K near room temperature. As the temperature rises, the C_p value increases gradually and begins to decrease, indicating an irreversible structural relaxation at 410 K. With a further increase in temperature, the C_p value shows its minimum at 570 K, then increases rapidly in the glass transition range from 600 to 670 K and reaches 47.0 J/mol·K for the supercooled liquid around 685 K. With further increased temperature, the C_p value of the supercooled liquid decreases gradually and then rapidly due to crystallization at 740 K. It is seen in Fig. 1.10 that the transition of the amorphous solid to the supercooled liquid takes place accompanied by a large increase in the specific heat, $\triangle C_{p,s \to l}$, reaching 14.5 J/mol·K. In addition, the $\triangle C_{p,s \to l}$ value for the $Zr_{65}Al_{7.5}Cu_{2.5}(Co,Ni,Cu)_{25}$ amorphous alloys was examined as a function of TM composition. The $\triangle C_{p,s \to l}$ value is in the range of 13 to 15 J/mol·K over the whole compositional range, and no distinct compositional dependence of the $\triangle C_{p,s \to 1}$ value is seen for the Zr-based alloys with $\triangle T_x$ values larger than about 30 K.

1.7 Physical Properties

1.7.1 Density

The anomalous X-ray scattering analyses have previously pointed out that satisfaction of the three empirical rules for achievng of high GFA causes the formation of a denser random packed structure in terms of topological and chemical atomic configuration. Furthermore, the short-range atomic configurations are also different from those for the corresponding crystalline alloys. These data allow us to conclude that the supercooled liquid in the multicomponent systems has a new atomic configuration which has not been obtained for other metallic alloys. It is expected that the unique atomic configuration changes the density of the new bulk amorphous alloys, though the density for previously reported amorphous alloys with low GFA decreases by about 2 % upon amorphization [2,3,46]. The densities in the as-cast, relaxed, and fully crystallized states have been measured for the Zr-Al-Cu, Zr-Al-Ni-Cu, and $Pd_{40}Cu_{30}Ni_{10}P_{20}$ amorphous alloys [46]. The densities increase steadily upon structural relaxation and crystallization. The increasing ratios of density defined by $(\rho_c - \rho_a)/\rho_a$ are as small as 0.30 to 0.5% much smaller than those (about 2%) [2,3,46] for previously reported amorphous alloys. The small ratios indicate that the bulk amorphous alloys have a denser random packing state of the constituent atoms. This result is consistent with the tendency obtained from structural analyses. In the higher packing density state, the atomic rearrangements of the constituent elements are suppressed, leading to decrease in atomic mobility and increase in viscosity. Furthermore, the frequency at which the atoms are trapped at the liquid/solid interface decreases in the denser random packing state, leading to decrease in the nucleation frequency. These changes in mobility and viscosity are interpreted causing an the increase in T_g/T_m and decrease in the growth rate of a crystalline phase.

1.7.2 Electrical Resistivity

Figure 1.11 shows the DSC thermogram and the corresponding electrical resistance curve, $R(T)/R_o$, around the supercooled liquid region for the $Zr_{60}Al_{15}Ni_{25}$ alloy [71], where R_o represents the electrical resistance at 300 K. The temperature derivative curve is also plotted because the onset of the glass transition is not obvious on the $R(T)$ curve. The T_g determined from the differential curve agrees with T_g on the DSC curve. The T_x also agrees with each other. Because T_g and T_x estimated from the $R(T)$ measurement agree with those of the DSC thermogram, the $R(T)$ measurement was also used to

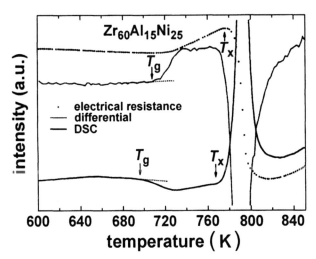

Fig. 1.11. DSC thermogram and electrical resistance curve along with its temperature derivative curve around the supercooled liquid region. The samples were prepared with v_s=40 m/s in an argon atmosphere. All measurements were carried out in an argon atmosphere

investigate the thermal stability of amorphous alloys. The R value around room temperature decreased with increasing T_a. It has been reported [72] that an amorphous $(Ni_{0.33}Zr_{0.67})_{85}Al_{15}$ alloy exhibits a negative temperature derivative of the resistivity in the temperature range from 3 to 300 K. Before the glass transition occurs, the R value increases temporarily, starting from 450 K, although the temperature derivative $d(R(T)/R_o)/dT$ still remains negative. This is responsible for the onset of structural relaxation [21]. In the supercooled liquid region, the $d(R(T)/R_o)/dT$ turns positive and is almost linear with temperature (see also Fig. 1.11). It is predicted that the electron transport property is significantly changed after the glass transition, although reliable experimental evidence except for electrical resistance and specific heat (DSC) is not presented. A similar $R(T)$ curve has been obtained for the $Zr_{60}Al_{15}Ni_{7.5}Co_{2.5}Cu_{15}$ alloy. The behavior of the $R(T)$ below T_g is similar to that of the $Zr_{60}Al_{15}Ni_{25}$ alloy. The $d(R(T)/R_o)/dT$ becomes positive in the supercooled liquid region, similarly. However, the $d(R(T)/R_o)/dT$ clearly changes in the two steps as shown in the inset, where the first step ranges from T_g to 765 K and the second ranges from 765 K to apparent T_x (775 K). After being heated to 761 and 771 K, the samples were quickly cooled to room temperature and examined by X-ray diffraction. The diffraction profile of the sample heated to 771 K shows the existence of some crystallites, whereas any Bragg peaks are invisible for the sample heated to 761 K. This suggests that crystallization starts from about 765 K. However, a definite exothermic peak cannot be observed in the DSC thermogram [73] around 765 K. Accordingly, the volume fraction of the crystallized region is presumed to be very small. The $R(T)$ in the supercooled liquid region of the $Pd_{40}Cu_{30}Ni_{10}P_{20}$ amorphous alloy was also examined [74]. Figure 1.12 shows the temperature dependence of normalized electrical resistivity ($\triangle R/R$) of a cast Pd-Cu-Ni-P amorphous sheet 1 mm thick and 5 mm wide. The resistivity decreases almost linearly in the temperature range up to 450 K and deviates slightly positively from the linear relationship in the range from 450 K to T_g. With further increasing temperature, the resistivity decreases significantly through two stages marked I and II in the supercooled liquid region and then more rapidly due to crystallization. Although the decrease in resistivity due to crystallization agrees with the previous data [75] for a number of amorphous alloys, the distinct two-stage reductions of resistivity in the supercooled liquid region are not always consistent with previous data [73] where a slight increase has been reported. The two-stage reductions in the resistivity in the supercooled liquid region suggest that the relaxation in the atomic configurations into the internal equilibrium state occurs through two stages and the relaxed atomic configurations leading to the decrease in resistivity also have two metastable states. The resistivity decreases from 2.3 to 1.5 µΩm with increasing Cu content from 0 to 40%, and the temperature coefficient of resistivity (TCR) changes from negative to positive passing through zero around 13 at % Cu and 1.8 µΩm. The relationship between the electrical resistivity at room temperature and the TCR for the $Pd_{40}Ni_{40-x}Cu_xP_{20}$ amorphous alloys indicates that the Mooij relationship [76] is also satisfied for the present Pd-Cu-Ni-P amorphous alloys.

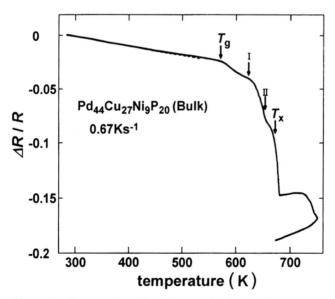

Fig. 1.12. Temperature dependence of relative electrical resistivity ($\triangle R/R$) for a cast amorphous $Pd_{44}Cu_{27}Ni_9P_{20}$ sheet 1 mm thick and 5 mm wide

1.7.3 Thermal Expansion Coefficient

Figure 1.13 shows the thermal dilatation ($\Delta L/L$) curves under a small tensile stress of 0.7 MPa for the melt-spun Pd-Cu-Ni-P amorphous ribbon, together with the data for T_g and T_x determined from the DSC curve at the same heating rate of 0.17 K/s. No appreciable change in $\Delta L/L$ is seen in the temperature range below 580 K, indicating that the alloy has an extremely low coefficient of thermal expansion in the amorphous solid. With further increasing temperature, the alloy begins to elongate at 580 K, followed by a rapid increase in elongation in the range of 600 to 615 K and then an instantaneous cessation of elongation around 620 K. Compared with the T_g and T_x values marked with arrows in Fig. 1.13, it is seen that the elongation starts at a temperature 30 K higher than T_g and then stops at a temperature 30 K lower than T_x. Considering that the onset temperature of elongation agrees with the temperature when the supercooled liquid reaches the truly internal equilibrium state, the extremely large elongation is obtained only in the supercooled liquid state. On the other hand, the elongation stops at 620 K in the supercooled liquid region. Because the mark of elongation is due to the start of crystallization, the difference is attributed to the acceleration of crystallization in the supercooled liquid caused by applying the small tensile load. The temperature coefficient of thermal expansion is determined to be 8×10^{-6} K^{-1} for the amorphous solid in the range below 500 K and 2.6×10^{-2} K^{-1} for the supercooled liquid in the range from 602 to 613 K. The change in the temperature coefficient of thermal expansion in the amorphous solid and supercooled liquid with Cu content was also examined for melt-spun $Pd_{40}Ni_{40-x}Cu_xP_{20}$ amorphous ribbons. The tensile load is in the narrow range of 0.7 to 1.3 MPa, and hence the influence of the difference in applied

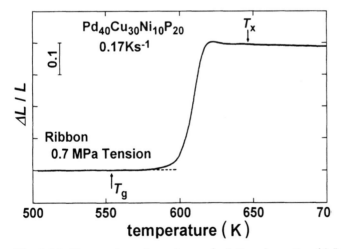

Fig. 1.13. Temperature dependence of relative elongation ($\Delta L/L$) under a tensile stress of 0.7 MPa for a melt-spun $Pd_{40}Cu_{30}Ni_{10}P_{20}$ amorphous ribbon

load on the temperature coefficient of thermal expansion can be neglected. The coefficient of thermal expansion for the amorphous solid in the range from room temperature to 500 K is 1.0×10^{-5} K^{-1} at 0 % Cu, shows a minimum value of 0.8×10^{-5} at 10 % Cu and then increases gradually to 1.8×10^{-5} K^{-1} with increasing Cu content to 40 % Cu. The coefficient is 1.3×10^{-5} K^{-1} for the $Pd_{40}Cu_{30}Ni_{10}P_{20}$ alloy with the largest GFA. The lowest coefficient (8×10^{-6} K^{-1}) is of the same order as that (1.2×10^{-6} K^{-1}) [77] for a typical invar Fe-36.5 mass % Ni alloy, indicating that the Pd-Cu-Ni-P amorphous alloys have very low coefficients of thermal expansion in metallic materials. On the other hand, the coefficient of thermal expansion for the supercooled liquid is as high as 3.4×10^{-2} K^{-1} at 0 % Cu.

1.8 Mechanical Properties

Cylindrical amorphous alloys prepared by copper mold casting are formed at different cooling rates in the inner and outer surface regions. The different cooling rates are expected to cause the difference in mechanical properties, though no appreciable difference is detected for T_g, T_x, and ΔT_x. We measured the Vickers hardness (H_v) as a function of the distance from the central point in the transverse cross section for cylindrical $Zr_{60}Al_{10}Co_3Ni_9Cu_{18}$ amorphous alloys with the diameters of 5 and 7 mm. The average H_v value was about 470 for the 5 mm cylinder and 475 for the 7 mm cylinder, agreeing with that (460) for melt-spun ribbon.

It is expected that the amorphous Zr-Al-M alloys prepared at cooling rates higher than the R_c will exhibit high tensile fracture strength (σ_f) comparable to that for melt-spun amorphous ribbons. The test specimen of the amorphous $Zr_{60}Al_{10}Co_3Ni_9Cu_{18}$ alloy was made by mechanical working from an as-cast amorphous cylinder 5 mm$^\phi$ × 50 mm. The gauge dimension has a diameter of 2 mm and a length of 10 mm as shown in Fig. 1.14a. The tensile stress-strain curves obtained for the two test specimens are also shown in Fig. 1.14b [78]. It is seen that the alloy is subjected to elastic deformation, followed by yielding, slight plastic deformation accompanying a slight work hardening, distinct plastic flow accompanying serration, and then final fracture. The feature of the stress-strain curve without significant work hardening and plastic elongation is in agreement with that for conventional amorphous metallic alloys with good bending ductility. Note that the distinct plastic flow accompanying serration takes place before final fracture. The observation of the serrated flow phenomenon during tensile deformation has hitherto been reported only for noble metal-based amorphous alloys of Pd- and Pt-based systems [2]. The serrated flow phenomenon implies that the shear deformation typical for ductile amorphous alloys does not occur catastrophically over the whole cross section and stops at the deformation site. The momentary interruption of shear deformation also implies the suppression of the final adiabatic fracture just after shear deformation, indicating

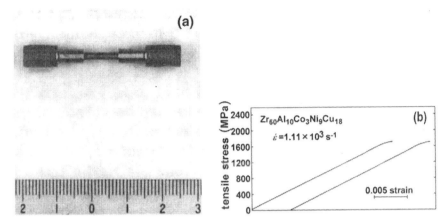

Fig. 1.14. (a) Outer surface appearance of an amorphous $Zr_{60}Al_{10}Co_3Ni_9Cu_{18}$ cylinder used for tensile testing, and (b) its tensile stress-strain curves at room temperature

that the bulk amorphous alloy is highly ductile comparable to that for noble-metal base amorphous alloys. Young's modulus (E), yield stress defined by the deviation from the proportional relationship, elastic elongation, plastic elongation, and σ_f are 97 GPa, 1390 MPa, 1.6 %, 0.4 % and 1510 MPa, respectively. The values of $\varepsilon_{y,t}(=\sigma_y/E)$, $\varepsilon_{y,c}(\equiv 3H_v/E)$ and σ_f/H_v are 0.14, 0.15 and 3.2, respectively. Furthermore, the ratio of σ_f to H_v indicates that the amorphous solid deforms in an ideal elastic-plastic mode without work hardening. These results indicate that the present bulk amorphous alloy has good ductility and exhibits tensile deformation and fracture behavior which agrees with that of an ideal elastic-plastic material. Figure 1.15 shows the tensile fracture surface appearance for a cast $Zr_{60}Al_{10}Co_3Ni_9Cu_{18}$ cylinder. The fracture occurs along the maximum shear plane which is declined by about 45 degrees in the direction of tensile load, and the fracture surface has a well-developed vein pattern. In comparison with the fracture surface

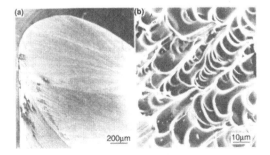

Fig. 1.15. Tensile fracture behavior (a) and fracture surface appearance (b) of an amorphous $Zr_{60}Al_{10}Co_3Ni_9Cu_{18}$ cylinder

for corresponding melt-spun amorphous ribbon, the diameter of their veins was about 1 to 2 μm for the cast cylinder, about 10 times as large as that (about 0.2 μm) for the melt-spun ribbon and comparable to that for the ribbon subjected to creep deformation at elevated temperatures near T_g. In addition, the SEM image shown in Fig. 1.15 reveals a remarkable development of the vein pattern caused by adiabatic deformation at the final fracture stage. The development of the vein pattern and the increase in the diameter of the veins suggest that the temperature during the final adiabatic fracture increases because final fracture resulting from good ductility is suppressed. Furthermore, the larger diameter of the veins also implies an increase in the thickness of the shear deformation region which induces an increase in the energy required for plastic deformation and final fracture. The compressive test method seems to be more appropriate for a more detailed investigation of serrated flow and work-hardening behavior. Figure 1.16 shows the compressive stress-strain curves for the cast $Zr_{60}Al_{10}Co_3Ni_9Cu_{18}$ cylinders of 2.5 mm$^\phi$ × 5.0 mm. Compared with those (Fig. 1.14) obtained by a tensile test, one can notice that the serration generates in a wider strain region from the yield point to the final fracture stage and no appreciable difference in the total strain up to final fracture is seen. Furthermore, the work-hardening accompanying the serrated flow becomes slightly distinct for the compressive test sample. Further development of serrated flow and the slight increase in the work-hardening presumably occur because the shear deformation bands generate along the entive outer surface in the compressive sample and the massive movement of the constituent atoms in the shear bands is suppressed by crossing among shear deformation bands. This presumption is supported the contrasting result that the shear deformation bands for the tensile sample generate along one direction. The inclination for generating shear deformation bands is presumably due to the difficulty of perfectly adjusting of the tensile axis in the tensile test because of scattering in the sample shape and the tensile axis in the tensile testing machine.

A Charpy test specimen with a U-shaped notch has a shape and dimension which agree with those for the JIS-No.3 (2.5 mm wide) type. The fracture

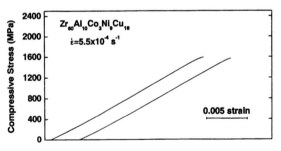

Fig. 1.16. Compressive stress-strain curves of amorphous $Zr_{60}Al_{10}Co_3Ni_9Cu_{18}$ cylinders

stress and displacement at impact fracture and the impact fracture energy were measured at room temperature with the Charpy testing machine [79]. The initial angle of the arm bar was 143.5°, and the moving velocity at the impact was 3.761 m/s. The load-displacement curve was stored in the computer, and the impact fracture energy was simultaneously calculated from the load-displacement curve. Figure 1.17 shows the Charpy impact load-displacement curves for $Zr_{55}Al_{10}Cu_{30}Ni_5$ and $Pd_{40}Cu_{30}Ni_{10}P_{20}$ amorphous alloys. The maximum load is measured as 4.3 kN for the Zr-based alloy and 4.4 kN for the Pd-based alloy, and hence the maximum impact fracture stress is determined to be 1615 and 1630 MPa, respectively. Note that the impact fracture stress is nearly the same as the static σ_f (about 1570 MPa [78] for the Zr-based alloy and 1670 MPa [56] for the Pd-based alloy) obtained in the wide strain rate range of 10^{-4} to 10^{-1} s^{-1} for the cast bulk amorphous alloys and the maximum displacement is as large as 0.44 to 0.48 mm. The impact fracture energy was 63 kJ/m^2 for the Zr-based alloy and 70 kJ/m^2 for the Pd-based alloy. Compared with the Charpy fracture energies for other metallic crystalline alloys, the present impact fracture energy largely exceeds those (12 to 36 kJ/m^2) [80] for newly developed Al-based high-strength alloys which have already been used in some applications. The comparison allows us to conclude that the present bulk amorphous alloys possess impact fracture energies which are high enough to enable their application as a structural material. The three peaks distinction in the impact load-displacement curve has frequently been observed in the Charpy test and are presumed to result from the harmonic vibrations of specimen at the impact fracture. It is important to clarify the impact fracture mode of the Zr- and Pd-based amorphous alloys with high impact fracture energies. The fracture surface consisted of the typical vein pattern in the region just near the U-shaped notch and changed to an equiaxed dimple-like pattern which occupies most of the region in the fracture surface. Neither a shell-like pattern nor a featureless cleavage-like

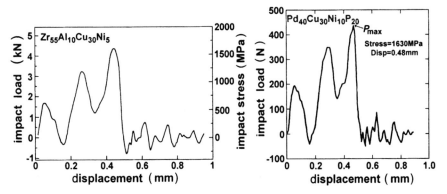

Fig. 1.17. Charpy impact fracture load displacement curves of the cast amorphous $Zr_{55}Al_{10}Cu_{30}Ni_5$ and $Pd_{40}Cu_{30}Ni_{10}P_{20}$ specimens

pattern typical of embrittled amorphous alloys was seen over the entire fracture surface. The features of the fracture surface indicate that the cast bulk amorphous alloy has good ductility leading to severe deformation even in the impact fracture mode. In addition, secondary failures were observed and a change in the appearance of the fracture surface from an equiaxed dimple-like pattern to a vein pattern occurred near the failures. The secondary failures appear to occur in close relationship to casting defects, that is, the casting defect corresponds to the interface between the formerly solidified amorphous region and the latterly solidified amorphous region. The amorphous phase region near the interface is subjected to subsequent heating. Heating decreases ductility compared with the as-cast amorphous phase without subsequent heating, leading to the generation of secondary failures. Although complete removal of the interface between formerly and latterly solidified amorphous alloys is difficult in the casting process, the decrease in the size of the interface is expected to increase further the impact fracture energy.

Based on the data described above, the mechanical properties of bulk amorphous alloys are summarized as follows: (1) higher σ_f and lower E compared with the corresponding crystalline phase, (2) much higher elastic strain of about 2% exceeding a yield limit of 0.2% for crystalline phases, (3) much higher elastic energy up to the yield point compared with crystalline alloys, (4) the absence of distinct plastic elongation at room temperature due to inhomogeneous deformation mode, and (5) relatively high impact fracture energy of about 70 kJ/m^2. When the stress-strain curves of Zr-based bulk amorphous alloys are compared with those for Ti-based crystalline alloys, the elastic energy required up to the yield point for a bulk amorphous alloy is larger by about 25 times. The relationships between E and σ_f or H_v for the bulk amorphous alloys are shown in Fig. 1.18, where the data of con-

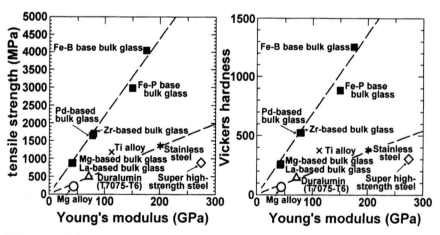

Fig. 1.18. Relationship between Young's modulus (E) and tensile strength (σ_f) or Vickers hardness (H_v) for bulk amorphous and crystalline alloys

ventional crystalline alloys are also plotted for comparison [81]. When the E values of bulk amorphous alloys are compared with those of crystalline alloys with the same σ_f, the moduli are smaller by about 60 % than those for crystalline alloys. The relationship between E and σ_f or H_v can be classified into two different groups. The distinct classification indicates that the mechanical properties of bulk amorphous alloys are different from those of crystalline alloys. The distinct difference in the fundamental mechanical properties is very important for applications. Some engineering applications based on the difference are in progress because of the unique characteristics which cannot be obtained from crystalline alloys in each application.

1.9 Viscoelasticity

Dynamic mechanical properties of storage E' and loss E'' module were measured with a dynamic mechanical analyzer by a forced oscillation method. Complex modulus E^* is the ratio of an oscillatory stress to an oscillatory strain. The real and imaginary parts of E^* are defined as E' and E'' and are described by equation (8) [82]

$$E^* = E' + iE'' = E\omega^2\tau^2/(1+\omega^2\tau^2) + iE\omega\tau/(1+\omega^2\tau^2). \tag{1.8}$$

Here, E is the elastic modulus, ω is the measurement frequency, and τ is the relaxation time. At $\omega\tau = 1$, E' and E'' show an inflection point of a reduction curve and a loss peak, respectively. The temperature-dependent measurements were carried out at every two degrees at heating rates from 0.0083 to 0.25 K/s at frequencies of 0.628, 6.28, and 62.8 rad/s. Figure 1.19 shows the temperature dependence of E' for the $La_{55}Al_{25}Ni_{20}$ amorphous alloy with heating rate α at 6.28 rad/s [83]. At 0.017 K/s, E' starts to reduce at 466 K due to the lower relaxation system. The effect of the higher temperature relaxation system appears in $E'(T)$ at 481 K as a plateau which indicates the transition of the main relaxation system in the measurement from the lower temperature relaxation to the higher temperature relaxation. This leads to a second rapid reduction of E' starting at 485 K with increasing temperature. From 503 K, E' increases due to a crystallization with further increasing temperature. At a higher α of 0.083 K/s, E' follows the same $E'(T)$ curve of the first reduction at 0.017 K/s, but E' shows more reduction superimposed on the first reduction curve at 0.017 K/s until 487 K. After the transition of the main relaxation system, E' reduces again due to the higher temperature relaxation and rejoins the $E'(T)$ curve at 0.017 K/s above 493 K. The appearance of the second relaxation system does not depend on the stress level but on temperature, suggesting the α dependence, that is, shifting to higher temperature with increase of α. E' shows more secondary reduction to a lower value compared with the $E'(T)$ curve at 0.017 K/s because the T_x shifts to a higher temperature of 517 K at 0.083 K/s. With further increasing α to 0.13 K/s, the change in the $E'(T)$ curve has the same tendencies. The

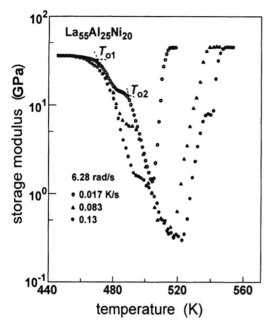

Fig. 1.19. Change in the temperature dependence of storage modulus (E') with heating rate at 6.28 rad/s

first and the secondary reduction $E'(T)$ curves are the same as the superimposed curves at lower α, but the appearance shifts to the higher temperature of 493 K. The T_x also shifts to 521 K.

The onset temperatures, T_{o1} and T_{o2}, for the first and the second reduction of E', respectively, are defined as the points of intersection of two tangent lines of the $E'(T)$ curve. The α dependence of T_{o1} and T_{o2} at the three frequencies, 62.8, 6.28, and 0.628 rad/s, is shown in Fig. 1.20. No different tendency is observed in the results for different frequencies. With increasing α, T_{o1} increases negligibly to a saturated value, whereas T_{o2} increases distinctly. For example, in the case of 0.628 rad/s, T_{o1} increases from 466 to 467 K, and T_{o2} increases from 481 to 493 K with increasing α from 0.0083 to 0.083 K/s. Because the appearance of the second relaxation in $E'(T)$ has α dependence, T_{o2} shows a distinct α dependence. The second relaxation starts to appear before the first relaxation shows a steep reduction in the $E'(T)$ curve at lower α (see Fig. 1.19). Thus, the tangent line of the reduction for T_{o1} becomes sluggish, leading to slight shift to a lower temperature for T_{o1}. At sufficiently higher α, as the first relaxation shows a steep reduction, T_{o1} shows very little α dependence and saturated value.

Figure 1.21 shows the temperature dependence of E'' of the La-Al-Ni alloy with α at 6.28 rad/s [83]. Two relaxation peaks for the lower and the higher temperature relaxation and the crystallization peak are clearly observed at 473, 490, and 511 K, respectively, at 0.017 K/s. The valley temperature of

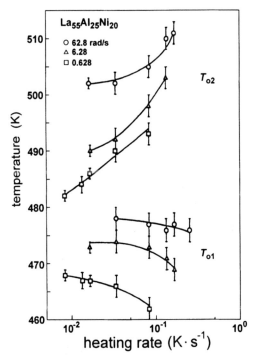

Fig. 1.20. Heating rate dependence of T_{o1} and T_{o2} at three frequencies, 0.628, 6.28, and 62.8 rad/s

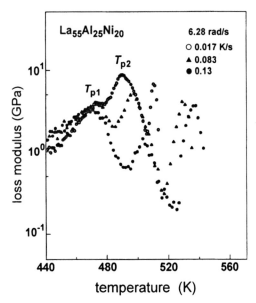

Fig. 1.21. Change in the temperature dependence of loss modulus (E'') with a heating rate at 6.28 rad/s

479 K between the two relaxation peaks corresponds to the temperature for the appearance of the effect of the second relaxation in the $E'(T)$ curve, as shown in Fig. 1.19. The peak temperatures, T_{p1} and T_{p2}, for the first and second relaxations, respectively, show contrary results for the increase in α. T_{p1} slightly shifts to lower temperatures from 473 to 471 K, and the height of the peak slightly decreases, whereas T_{p2} clearly shifts to higher temperatures from 490 to 501 K and the height rapidly decreases with increasing α from 0.017 to 0.13 K/s. At the higher α, the appearance of the effect of the second relaxation shifts to higher temperature marking the shift of T_{p2}, and the overlap of the two relaxation peaks becomes loose. Thus, the first peak appears distinctly with a slight shift to a lower temperature and a slight decrease in height. Because the initial E'' value for the second relaxation becomes lower due to the movement of the first peak, the height of the second peak decreases rapidly with increasing α. The $E''(T)$ curve merges with that of a lower α at temperatures above T_{p2} showing the same tendency as seen in the $E'(T)$ curve. The shifting behavior of T_{p1} and T_{p2} against α at three frequencies is summarized in Fig. 1.22. As the two relaxation peaks become more distinct at the lower frequency [84], T_{p1} and T_{p2} at the lower frequency shift more drastically than those at the higher frequency.

The endothermic reaction for the glass transition in the DSC curve also shifts to a higher temperature with increasing α. To correlate the change in

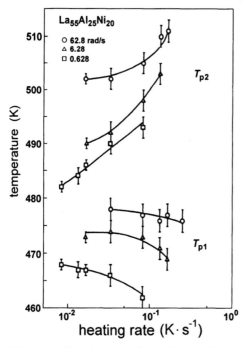

Fig. 1.22. Heating rate dependence of T_{p1} and T_{p2} at three frequencies

$E''(T)$ with α, apparent activation energies Q are determined by the Kissinger equation [85]. Figure 1.23 shows the Kissinger plots for T_{p2} and T_g. Here, we define T_g as an inflection point of an endothermic reaction to compare it with T_{p2} of the peak temperature. Peak temperatures for the crystallization, T_x, by DSC and $E''(T)$ curves are also shown in the figure (solid symbols), and these Q_s agree well with the value of about 170 kJ/mol. This confirms the validity of this method of comparing the data from $E''(T)$ with those from DSC by Q. Q_s obtained from the T_{p2}s at the three frequencies vary slightly from 300 to 340 kJ/mol which agree with the value of 340 kJ/mol from the DSC measurements. It should be mentioned that Q of about 340 kJ/mol represents the value for the relaxation and also the value including the effects of phase separation and relaxation. The real activation energy for the higher temperature relaxation has been obtained as 550 kJ/mol [84]. This result reveals that the shift of the DSC curve is affected mainly by the higher temperature relaxation. Comparing the ratio of E'' to E', $\tan\delta(=E''/E')$ between the two relaxations, $\tan\delta$ for the higher temperature relaxation exceeds 10°, is one order higher than that for the lower temperature relaxation [86]. It suggests that the higher temperature relaxation occurs more drastically than the lower temperature relaxation. The appearance of the α dependence of T_g is

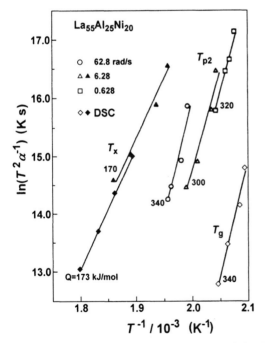

Fig. 1.23. Kissinger plots for T_g obtained by DSC and T_{p2} at three frequencies (open symbols). Solid symbols represent the data for the crystallization peaks from DSC and E'' at 6.28 rad/s. α is the heating rate (K/s)

caused mainly by phase separation, as evidenced by two E'' relaxation peaks for the $La_{55}Al_{25}Ni_{20}$ amorphous alloy. To examine the α dependence of the normal glass transition without the effect of phase separation, polymethyl methacrylate, PMMA [87] is used. As amorphous alloys with one relaxation peak, to our knowledge, do not have a stable supercooled liquid region, the crystallization peak of E'' overlaps the relaxation peak. The α dependence is hardly observed in mechanical measurements because of the margin of errors. Almost all amorphous alloys with a wide supercooled liquid region exceeding 40 K in the DSC curve at 0.67 K/s, such as $Zr_{60}Al_{15}Ni_{25}$, $Mg_{65}Cu_{25}Y_{10}$, and $Pd_{48}Ni_{32}P_{20}$ amorphous alloys, have two relaxation systems suggesting phase separation in the glass transition region [88]. These alloys show distinct α dependence of T_g, like the $La_{55}Al_{25}Ni_{20}$ alloy. On the other hand, amorphous alloys with one relaxation system and a smaller supercooled liquid region, such as $Zr_{70}Cu_{30}$, $La_{55}Al_{45}$, and $Pd_{80}Si_{20}$ amorphous alloys, show little α dependence of T_g. A similar two-stage change in dynamic mechanical properties has been seen for the $Zr_{60}Al_{15}Ni_{25}$ amorphous alloy [89].

1.10 Soft Magnetic Properties

According to the three empirical rules for achieving high GFA, a new bulk amorphous alloy with ferromagnetism at room temperature has been investigated. As a result, as described in Sect. 1.1, soft ferromagnetic bulk amorphous alloys have been found in Fe-(Al,Ga)-(P,C,B,Si) [27,28,90], Co-Cr-(Al,Ga)-(P,B,Si) [91], Fe-(Co,Ni)-(Zr,Nb,Ta)-B [29–31,92], and Co-Fe-(Zr,Nb)-B [32] systems. This section describes the soft magnetic properties of Fe- and Co-based bulk amorphous alloys.

1.10.1 Formation and Soft Magnetic Properties of Bulk Amorphous Alloys

Fe-based amorphous alloys with large ΔT_x above 60 K are expected to have high GFA which enables the production of bulk amorphous alloys with diameters above 1 mm by casting processes. Cast $Fe_{72}Al_5Ga_2P_{11}C_6B_4$ [27] and $Fe_{72}Al_5Ga_2P_{10}C_6B_4Si_1$ [90,93] amorphous cylinders with diameters of 1 to 2 mm had smooth surfaces and metallic luster. In addition, agreement of the outer shape of the cast samples with the inner cavity of the copper mold indicates good castability for these Fe-based alloys. Even after an appropriate etching treatment, no appreciable contrast corresponding to a crystalline phase is seen over the whole transverse cross section of the cast $Fe_{72}Al_5Ga_2P_{10}C_6B_4Si_1$ cylinders with diameters of 1 and 2 mm. In addition, neither cavities nor shrinkage holes are observed, indicating that the amorphous alloy has good castability due to the lack of discontinuous change in the specific volume-temperature relation which ordinarily occurs in crystalline alloys obtained from liquid. The cast $Fe_{72}Al_5Ga_2P_{11}C_6B_4$ amorphous cylinders with diameters of 0.5 to 1.5 mm [27] show the sequential changes of the glass transition, supercooled liquid, and then single-stage exothermic peaks which agree with those of the corresponding melt-spun amorphous ribbon. No distinct differences in the thermal stability and crystallization mode of the supercooled liquid are seen between cast amorphous cylinders and melt-spun amorphous ribbon.

The hysteresis $B-H$ loop of a cast $Fe_{72}Al_5Ga_2P_{11}C_6B_4$ amorphous cylinder with a diameter of 1 mm was examined in the as-cast and annealed (723 K, 600 s) states [27,90]. The B_s, H_c and B_r/B_s of the annealed sample are 1.07 T, 5.1 A/m and 0.37, respectively, indicating that the cast amorphous cylinder has good soft magnetic properties. In addition, the μ_e at 1 kHz for the annealed amorphous cylinder also has a high value of 7000. The soft magnetic properties were further improved for the 1 at % Si-containing alloy [90,93]. Figure 1.24 shows the hysteresis $B-H$ loop for a cast $Fe_{72}Al_5Ga_2P_{10}C_6B_4Si_1$ amorphous cylinder with a diameter of 2 mm, together with the data for a cast $Fe_{73}Al_5Ga_2P_{11}C_5B_4$ amorphous cylinder with a diameter of 1 mm. The B_s, H_c and B_r/B_s for the Si-containing cylinder are 1.14 T, 0.5 A/m and 0.38, respectively. In comparison with the values [27] for the Fe-Al-Ga-P-C-B cylinder, H_c decreases and B_r/B_s increases for the Si-containing amorphous cylinder, though B_s decreases slightly due to the slightly lower Fe concentration. The thermomagnetic data confirmed that cast $Fe_{73}Al_5Ga_2P_{11}C_5B_4$ amorphous cylinder has a Curie temperature (T_c) at 600 K which is lower by 185 K than the T_x. The T_c value agrees with that determined from the endothermic peak on the DSC curve. It is thus concluded that the cast Fe-based

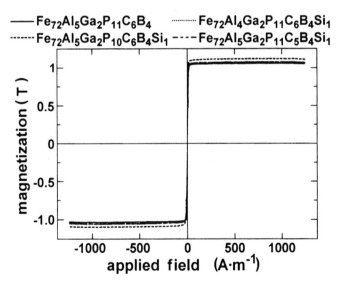

Fig. 1.24. Hysteresis $B - H$ loops of cast amorphous Fe-Al-Ga-P-C-B and Fe-Al-Ga-P-C-B-Si cylinders

amorphous cylinders with diameters up to 2 mm exhibit good soft magnetic properties of 1.1 T for B_s, 2 to 6 A/m for H_c and 7000 for μ_e at 1 kHz. When these soft magnetic properties are compared with those for corresponding melt-spun amorphous ribbons, no distinct changes in B_s and H_c are seen. However, the μ_e value is degraded for the cast cylinder presumably because of the increasing influence of demagnetization resulting from the significant change in the sample morphology.

Figure 1.25 shows the outer morphology of bulk $Fe_{61}Co_7Zr_{10}Mo_5W_2B_{15}$ cylinders with diameters of 3 and 5 mm [31]. These samples also have a smooth surface and metallic luster, and no contrast of a crystalline phase is seen over the outer surface. The X-ray diffraction patterns showed a main halo peak with a wave vector $K_p(=4\pi\sin\theta/\lambda)$ around 29.6 nm^{-1} and no crystalline peak is observed even for the 5 mm$^\phi$ sample. In addition, the optical micrographs of the cross section of the two samples also revealed a featureless contrast in an etched state using hydrofluoric acid. These results indicate that the bulk cylinders are composed of an amorphous phase in the diameter range up to 5 mm. Considering that the bulk cylinder of 7 mm in diameter consists of an amorphous phase in an outer surface region about 2 mm thick and amorphous and crystalline phases in the inner region, the t_{max} for the $Fe_{61}Co_7Zr_{10}Mo_5W_2B_{15}$ alloy is determined to be about 6 mm. The t_{max} is 3 times larger than the largest value (2 mm for $Fe_{72}Al_5Ga_2P_{10}C_6B_4Si_1$) [90,93] for Fe-based amorphous alloys reported to date. Figure 1.26 shows the DSC curves of bulk amorphous $Fe_{61}Co_7Ni_7Zr_8Nb_2B_{15}$, $Fe_{56}Co_7Ni_7Zr_8Ta_2B_{20}$, $Fe_{60}Co_8Zr_8Nb_2Mo_5W_2B_{15}$, and $Fe_{62}Co_8Zr_8Mo_5W_2B_{15}$ cylinders with diameters of 1 to 3 mm [31]. These amorphous alloys exhibit the sequential transition of glass transition, supecooled liquid, and crystallization. The ΔT_x is as large as 55 to 88 K, and crystallization occurs by a single exothermic reaction. The crystallites consist of α-Fe, Fe_2Zr, Fe_3B, MoB, and W_2B phases for the $Fe_{60}Co_8Zr_{10}Mo_5W_2B_{15}$ sample heated to the temperature just above the exothermic peak. Thus, crystallization is due to the simultaneous precipitation of the crystalline phases. This crystallization mode agrees with

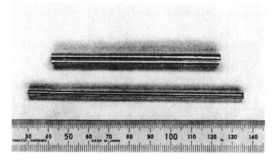

Fig. 1.25. Outer morphology of cast amorphous $Fe_{61}Co_7Zr_{10}Mo_5W_2B_{15}$ cylinders with diameters of 3 and 5 mm

that [37–41] for other bulk amorphous alloys. The largest ΔT_x is 88 K for $Fe_{56}Co_7Ni_7Zr_8Ta_2B_{20}$, larger than the largest values (57 to 67 K) for Fe-(Al,Ga)-(P,C,B,Si) [27,93] and nonferrous Pd- and Pt-based amorphous alloys [2,15,16]. The T_m was 1420 K for $Fe_{56}Co_7Ni_7Zr_{10}B_{20}$ and 1416 K for $Fe_{61}Co_7Zr_{10}Mo_5W_2B_{15}$ and the T_g/T_m was determined to be 0.60 for the former alloy and 0.63 for the latter alloy. Considering that T_g/T_m is 0.54 for $Fe_{80}P_{12}B_4Si_4$ [94] and 0.57 for $Fe_{73}Al_5Ga_2P_{11}C_4B_4Si_1$ [93], the present T_g/T_m values are believed to be the highest among all Fe-based amorphous alloys.

Table 1.4 summarizes the t_{max}, T_g, ΔT_x, T_g/T_m, compressive fracture strength ($\sigma_{c,f}$), H_v, B_s, H_c, μ_e at 1 kHz, and λ_s for the new amorphous Fe-(Co,Ni)-(Zr,Nb,Ta)-B [29,30,92] and Fe-Co-Zr-(Mo,W)-B alloys [31]. These amorphous alloys exhibit good soft magnetic properties in an annealed (800 K, 300 s) state, i.e., high B_s of 0.74 to 0.96 T, low H_c of 1.1 to 3.2 A/m, high μ_e of 12000 to 25000, and low λ_s of 10–14 × 10^{-6}. The H_c and μ_e are superior to those for conventional Fe-Si-B amorphous ribbons [95–97], presumably because of the lower λ_s. Furthermore, the $Fe_{60}Co_8Zr_{10}Mo_5W_2B_{15}$ bulk amor-

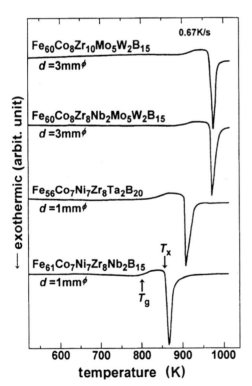

Fig. 1.26. DSC curves of cast amorphous $Fe_{61}Co_7Ni_7Zr_8Nb_2B_{15}$, $Fe_{56}Co_7Ni_7Zr_8Ta_2B_{20}$, $Fe_{60}Co_8Zr_8Nb_2Mo_5W_2B_{15}$, and $Fe_{60}Co_8Zr_{10}Mo_5W_2B_{15}$ cylinders

Table 1.4. Thermal stability, mechanical strength, and magnetic properties of cast bulk amorphous Fe-(Al,Ga)-(P,C,B,Si), Fe-(Co,Ni)-(Zr,Nb,Ta)-B, and Fe-Co-Zr-(Mo,W)-B alloys

Alloy	Thermal stability			Mechanical strength		Soft magnetic properties			
	T_g (K)	ΔT_x (K)	T_g/T_m	H_v	$\sigma_{c,f}$ (MPa)	B_s (T)	H_c (A/m)	μ_e (1kHz)	λ_s (10^{-6})
$Fe_{56}Co_7Ni_7Zr_{10}B_{20}$	814	73	0.60	1370	—	0.96	2.0	19100	10
$Fe_{56}Co_7Ni_7Zr_8Nb_2B_{20}$	828	86	—	1370	—	0.75	1.1	25000	13
$Fe_{61}Co_7Ni_7Zr_8Nb_2B_{15}$	808	50	—	1340	—	0.85	3.2	12000	14
$Fe_{56}Co_7Ni_7Zr_8Ta_2B_{20}$	827	88	—	1360	—	0.74	2.6	12000	14
$Fe_{60}Co_8Zr_{10}Mo_5W_2B_{15}$	898	64	0.63	1360	3800	—	—	—	14

phous alloy has high $\sigma_{c,f}$ and H_v of 3800 MPa and 1360, respectively, which greatly exceed those (σ_f=1000-2000 MPa, H_v=300-800) [98] for high-carbon, high-alloy tool steels and 25 mass % Ni maraging steel. No bulk alloys with high strength above 3000 MPa for $\sigma_{c,f}$ and 1000 for H_v have been obtained for any kinds of Fe-based alloys including those with amorphous and crystalline phases. In addition, no weight loss is detected after immersion for 3.6 ks at 298 K in aqua regia. Thus, the Fe-based bulk amorphous alloys possess simultaneously high GFA, high strength, high corrosion resistance, and good soft magnetic properties which cannot be obtained from other amorphous and crystalline alloys.

1.10.2 Glass-Forming Ability of Fe-(Al,Ga)-Metalloid, Fe-TM-B, and Co-TM-B Alloys

It is important to investigate the GFA of the new Fe-based amorphous alloys in comparison with other nonferrous metal base amorphous alloys. In Fig. 1.27, we plot the R_c, t_{max}, and T_g/T_m or ΔT_x of the Fe-(Al,Ga)-(P,C,B,Si), Fe-(Co,Ni)-(Zr,Nb,Ta)-B and Fe-Co-(Zr,Nb)-(Mo,W)-B amorphous alloys [31] in the relationships among those for typical bulk amorphous alloys reported to date. The t_{max}, T_g/T_m, and ΔT_x values of the present Fe-based amorphous alloys lie well in the previous empirical relationships. There is a clear tendency for t_{max} to increase with increasing T_g/T_m and ΔT_x. The high GFA of the new Fe-based alloys is due to the same mechanism as that for other bulk amorphous alloys. The good fit predicts that the R_c is about 1000 K/s for the $Fe_{72}Al_5Ga_2P_{11}C_6B_4$ alloy, 300 K/s for the $Fe_{72}Al_5Ga_2P_{10}C_6B_4Si_1$ and $Fe_{72}Al_5Ga_2P_{11}C_5B_4Si_1$ alloys, 700 K/s for $Fe_{56}Co_7Ni_7Zr_{10}B_{20}$ and $Fe_{56}Co_7Ni_7Zr_8Nb_2B_{20}$, and 60 K/s for $Fe_{61}Co_7Zr_{10}Mo_5W_2B_{15}$.

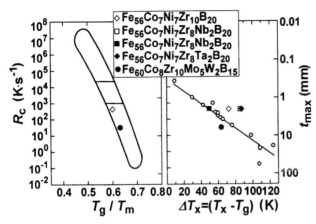

Fig. 1.27. Relationship among R_c, t_{max} and T_g/T_m or ΔT_x for the new Fe-based bulk amorphous alloys. The data for nonferrous metal-based bulk amorphous alloys are also shown for comparison

1.11 Viscous Flow and Microformability of Supercooled Liquids

In addition to the low R_c and large t_{max}, bulk amorphous alloys have an important feature in that a wide supercooled liquid region is obtained in the temperature range before crystallization [37–41]. The ΔT_x reaches as much as 127 K [70] for Zr-Al-Ni-Cu alloys and 98 K [51] for Pd-Cu-Ni-P alloys. In the supercooled liquid region with such large ΔT_x values, the viscosity decreases significantly, and high viscous flow is obtained [99]. By utilizing the low viscosity and high viscous flow, bulk amorphous alloys can be deformed to various complicated shapes and maintain good mechanical properties [100–103]. This section intends to describe the microformability in the supercooled liquid region for La-Al-Ni, Zr-Al-Ni-Cu and Pd-Cu-Ni-P bulk amorphous alloys and to introduce some practical examples of micro-formed bulk amorphous alloys.

1.11.1 Phase Transition of Bulk Amorphous Alloys

The phase transition of bulk amorphous alloys heated at a scanning rate of 1 K/s differs from that of conventional amorphous alloys which require high cooling rates above 10^4 K/s. Bulk amorphous alloys show the sequential transition of amorphous → glass transition → supercooled liquid region → crystallization upon continuous heating, whereas neither a glass transition nor a supercooled liquid region is observed for conventional amorphous alloys. Thus the bulk amorphous alloys have a broad supercooled liquid region before crystallization. The appearance of the broad supercooled liquid region also implies that the supercooled liquid has high resistance to crystallization. High thermal stability is the basis for achieving high GFA. The reason for the high stability of the supercooled liquid in the new systems has been described [37]-[41] in section 2 on the basis of the formation of a denser random packed structure in multicomponent alloys that satisfy the three empirical rules.

1.11.2 Deformation Behavior of Supercooled Liquids

The T_g and ΔT_x values of bulk amorphous $La_{55}Al_{25}Ni_{20}$, $Zr_{65}Al_{10}Ni_{10}Cu_{15}$ and $Pd_{40}Cu_{30}Ni_{10}P_{20}$ alloys measured at a heating rate of 0.67 K/s were, respectively, 480 and 70 K for the La-based alloy, 650 and 85 K for the Zr-based alloy, and 575 and 95 K for the Pd-based alloy. Figure 1.28 shows the relationship between the maximum flow stress (σ_{max}) and the strain rate ($\dot{\varepsilon}$) in the supercooled liquid for $La_{55}Al_{25}Ni_{20}$ and $Zr_{65}Al_{10}Ni_{10}Cu_{15}$ amorphous alloys [99,103,104]. The linear relationship holds at all deformation temperatures, and the slope corresponds to the strain-rate sensitivity exponent (m-value). As seen in the figure, the m-value is measured to be approximately 1.0, indicating that the supercooled liquid has an ideal Newtonian flow, i.e., ideal superplasticity. Figure 1.28 also shows that the deformation in the supercooled liquid takes place in a homogeneous mode over the whole strain rate range from 1.5×10^{-4} to $7 \times 10^{-1} s^{-1}$ [104]. This result also indicates that high strain-rate superplasticity is obtained in the supercooled liquid. The amorphous alloy exhibits an inhomogeneous deformation mode in the temperature range below T_g, and the flow stress decreases with increasing strain rate. Thus, deformation behavior differs significantly between the amorphous solid and supercooled liquid. Figure 1.29 shows the change in the viscosity of the supercooled liquid for the $Pd_{40}Cu_{30}Ni_{10}P_{20}$ amorphous alloy during isothermal heating [105]. The heating up to each deformation temperature was done at a heating rate of 0.33 K/s. Considering that the flow deformation becomes extremely easy in the viscosity range below 10^8 Pa·s, viscous

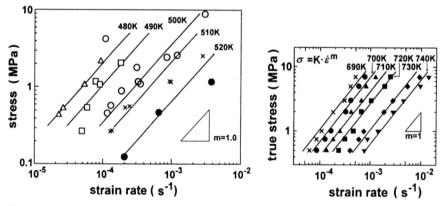

Fig. 1.28. Relationship between the maximum true flow stress (σ_{max}) and strain rate ($\dot{\varepsilon}$) in the supercooled liquid for amorphous $La_{55}Al_{25}Ni_{20}$ and $Zr_{65}Al_{10}Ni_{10}Cu_{15}$ alloys

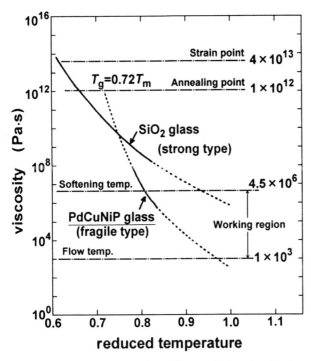

Fig. 1.29. Temperature dependence of viscosity in the supercooled liquid for the amorphous $Pd_{40}Cu_{30}Ni_{10}P_{20}$ alloy. The solid circles represent the experimental data and the three lines denote the calculated values, assuming different viscosities of the liquid at T_m

flow deformation can be done during long time period of 10^2 to 10^4 s in the temperature range between 600 and 650 K. The times are long enough to transform the viscous flow deformation into various kinds of complicated shapes because these amorphous alloys have high strain-rate superplasticity.

1.11.3 Microforming of Supercooled Liquids

By utilizing ideal superplasticity which can be achieved over the wide strain-rate range in the supercooled liquid region, various kinds of micro-forming treatments have been done for La-, Zr- and Pd-based amorphous alloys. Figure 1.30 shows the outer shapes of the $La_{55}Al_{25}Ni_{20}$ amorphous wire obtained by tensile deformation at an initial strain rate of about $1 \times 10^{-2} s^{-1}$ and 500 K [106] and La-based amorphous gear with an outer diameter of 1 mm prepared by die forging into gear-shaped silicon die for 10^3 s at 500 K [99]. We have also obtained [103] small amorphous gears by a die extrusion process in the supercooled liquid region for amorphous $Zr_{65}Al_{10}Ni_{10}Cu_{15}$ alloy with much higher T_g compared with that for La-based amorphous alloys.

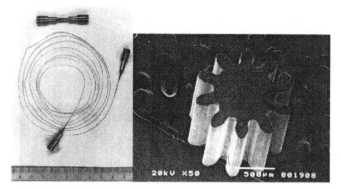

Fig. 1.30. Wire- and gear-shaped amorphous $La_{55}Al_{25}Ni_{20}$ samples prepared by tensile stretching and die forging treatments, respectively, in the supercooled liquid region

We examined the relationship between extrusion pressure (P_e) and extrusion velocity (V_e) at 711 K in the supercooled liquid region for the Zr-Al-Ni-Cu amorphous alloy. There is a good linear relationship and the m-value is determined to be 1.0, indicating that the supercooled liquid also exhibits ideal superplastic behavior even in die extrusion. Amorphous Zr-based alloy gears were also produced by the die extrusion method. The extruded sample does not contain any crystallinity and has the same T_g and T_x values as those for as-cast bulk amorphous alloys. Figure 1.31 shows an outer morphology of a precision optical mirror prepared by press forging the $Zr_{60}Al_{10}Ni_{10}Cu_{20}$ amorphous alloy for 600 s at 673K [107]. In the as-forged state, the forged sample exhibits a smooth outer surface and good metallic luster. The outer surface retains good smoothness on a nanometer scale based on the data obtained with a surface roughness indicator. The three-stage amorphous gear with outer diameters of 5, 6, and 7 mm was also prepared by press die-forging the $Pd_{40}Cu_{30}Ni_{10}P_{20}$ amorphous alloy for 600 s at 630 K [105]. All

Fig. 1.31. Fine precision amorphous alloy mirrors prepared by die forging an amorphous $Zr_{60}Al_{10}Ni_{10}Cu_{20}$ alloy in the supercooled liquid region

microformed alloys retain an amorphous phase, and no distinct differences in thermal stability and mechanical properties are seen between the as-cast and microformed samples [100–103]. The combination of high GFA, good microformability and good mechanical properties has already enabled some practical uses for these bulk amorphous alloys as microforged materials, such as fine machinery parts and fine precision optical parts.

1.12 Bulk Amorphous Alloys Produced by Powder Consolidation

By using high viscous flow and atomic diffusivity in the supercooled liquid, it is expected that a bulk amorphous alloy with full density can be produced from atomized amorphous powders. This production method causes another route for synthesizing a bulk amorphous alloy and has the great advantage of producing large scale bulk amorphous alloys independent of GFA. Great effort has been devoted to producing a bulk amorphous alloy by the powder consolidation method [108]. However, there have been no successful data on forming a bulk amorphous alloy with the same tensile strength as that of the corresponding melt-spun amorphous ribbon. The difficulty is probably due to the use of amorphous alloys without a glass transition and with a narrow supercooled liquid region below 30 K before crystallization. Significant extension of the supercooled liquid region has enabled the production of bulk amorphous alloys with full density and the same high σ_f as that for melt-spun amorphous alloy ribbons [109–111]. This section describes the fabrication of Zr-based bulk amorphous alloys by powder consolidation and their thermal stability and mechanical properties.

1.12.1 Consolidation Conditions

Atomized Zr-Al-Ni-Cu amorphous powder was consolidated via closed P/M processing. The powder was contained in a well-controlled atmosphere. The sieved powder with a particle size < 75 µm was cold-pressed to a density of about 65% of theoretical in a copper can 20 mm in inner diameter and 23 mm in outer diameter. Subsequently, the precompacted powder was degassed at 573 K for 900 s under 1×10^{-3} Pa, and then sealed by welding. The sealed powder which constituted the billet for extrusion or hot pressing was heated to temperature and then consolidated. The consolidated billet was quenched immediately in water. The billets were extruded at a uniform ram speed V_e and various extrusion ratios R using conical dies with a half die angle α of 30° and an extrusion container with an inner diameter D of 24 mm. The billets were also hot-pressed under 1.0 GPa for 180 s.

It is important to control the processing temperature and time to retain the single amorphous phase. The rise in temperature ΔT_e of the billet during extrusion, generated by work-induced heat, is given by [112]

$$\Delta T_e = 1.1 \times 10^4 \times V_e^{0.64} \times P_e/(\rho C_p), \tag{1.9}$$

where V_e, P_e, ρ, and C_p are the ram speed, pressure during extrusion, density, and specific heat of the alloy, respectively. The measured density ρ of the Zr-Al-Ni-Cu amorphous alloy was 6.70 g/cm^3 and C_p was estimated at 30 J/mol^{-1}K^{-1}. For V_e=1.0 mms^{-1} and a critical P_e of 1.0 GPa for the die assembly, the maximum ΔT_e was estimated at 55 K. The calculated optimum extrusion temperature for producing ductile amorphous compacts was >673 K; this is obtained by subtracting ΔT_e from the maximum embrittlement-proof temperature T_d of 728 K. Thus, we chose temperatures of 673 K, 693 K, and 713 K as the extrusion temperatures T_e for subsequent consolidation.

Consolidation must occur in the homogeneous mode of deformation. The deformation mode depends on the strain rate as well as the temperature. The average strain rate $\dot{\varepsilon}$ of the billet during extrusion is given by [113]

$$\dot{\varepsilon} = 6V_e \ln(R) \tan(\alpha(1/D)), \tag{1.10}$$

where V_e, R, D, and α are the extrusion speed (ram speed), extrusion ratio, inner diameter of the container, and the half angle of the die, respectively. D and α of the die assembly used in this study are 24 mm and 30°, respectively. The maximum strain rate for homogeneous deformation of the Zr-Al-Ni-Cu amorphous alloy at the minimum extrusion temperature of 673K is at least 0.5 s^{-1}. Under the most severe extrusion conditions (maximum R of 5 and minimum T_e of 673 K), the maximum ram speed V_e for extruding the Zr-Al-Ni-Cu amorphous powder homogeneously is estimated at about 2.0 mm/s. Thus, a ram speed of 1.0 mm/s was utilized.

1.12.2 Density and Properties of Consolidated Bulk Amorphous Alloys

Figure 1.32 shows the relationship between the extrusion ratio R, the extrusion temperature T_e and the structure of the extruded Zr-Al-Ni-Cu alloy together with the values of T_g, T_d and T_x. An amorphous phase is retained when T_e is below 693 K and $R \leq 4$, and < 673 K and R=5. Further increase in T_e results in the coexistence of a crystalline phase. X-ray diffraction patterns and DSC curves of bulk alloys consolidated at T_e=673 K show clearly that an amorphous phase (without crystallinity) is retained and the heat of crystallization ΔH_x is the same as that for the as-atomized powder. Figure 1.32 also shows bright-field electron micrographs and selected-area electron diffraction patterns of bulk alloys extruded at T_e=673 K and R=5, T_e=693 K and R=5, and T_e=693 K and R=4. It is seen that the extruded bulk alloys consist of a single amorphous phase under the first two conditions and amorphous and crystalline phases under the third condition. The structural and

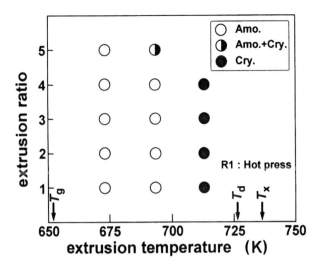

Fig. 1.32. Effect of extrusion temperature and extrusion ratio on the structure of $Zr_{65}Al_{10}Ni_{10}Cu_{15}$ compacts

thermal data indicate clearly that the consolidated bulk alloys represented by open symbols in Fig. 1.32 retain an amorphous structure. The relative density of the consolidated bulk alloys was measured as a function of extrusion temperature and extrusion ratio. A density above 99.5% of theoretical was achieved for all of the compacts consolidated at 673 K and 693 K. The density of the compacts hot-pressed at 713 K under 1.0 GPa was as low as 90% of theoretical because of the increase of flow stress due to crystallization. In the scanning electron micrograph (SEM) of a polished cross-section of the amorphous compact produced at $R=5$ and $T_e=673$ K, neither cavities nor pores are observed in the compact. Thus is clear that the compacts with a single amorphous phase and a density of more than 99.5% of theoretical were obtained over a wide range of extrusion conditions for the amorphous alloy with a broad supercooled liquid region. Tensile test specimens were spark machined and polished. The thickness, width and gauge length of the test pieces were 1.5, 3.0 and 8.4 mm, respectively. Tensile tests were conducted using an Instron testing machine at room temperature and at a strain rate of 5×10^{-4} s^{-1}. For comparison, the $Zr_{65}Al_{10}Ni_{10}Cu_{15}$ amorphous bulk alloy (with a transverse cross section of 1.5×5.0 mm^2) produced by injection casting of the melt into a copper mold was also tested using a test piece with the same dimensions as the compacts. Figure 1.33 shows the dependence of the tensile strength on the extrusion ratio R for compacts produced at extrusion temperatures of 673 K and 693 K via closed P/M processing. Hot-pressing is represented by $R=1$. Strength increases with increasing extrusion ratio and temperature. The bulk amorphous alloys produced at extrusion ratios >3 exhibit strength levels similar to those of the as cast bulk amorphous alloy (1570

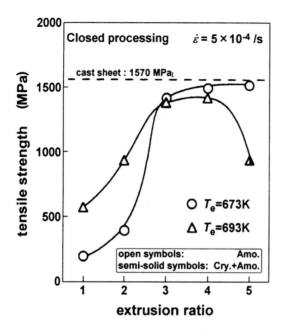

Fig. 1.33. Dependence of tensile strength on extrusion ratio for $Zr_{65}Al_{10}Ni_{10}Cu_{15}$ amorphous compacts

MPa) and melt-spun amorphous ribbon (1440 MPa). The Young's modulus of the compacts with a single amorphous phase was about 80 GPa, identical to that of the cast amorphous bulk alloy. The partially crystallized compact produced at T_e=693 K and R=5 has a much lower strength of 930 MPa, and the Young's modulus is a high value at 117 GPa. Figure 1.34 shows the tensile fracture surface of the amorphous alloy compact produced at T_e=673 K and R=5 which has the highest σ_f of 1520 MPa. The fracture takes place along the maximum shear plane which is inclined at 45° to the direction of tensile load. No fracture occurs along the boundaries between prior powder particles, which is attributed to good bonding between the particles. The high σ_f appears to originate from good powder particle bonding. Vein patterns, a typical fracture characteristic of amorphous alloys with good ductility, are also observed over the entire fracture surfaces. No shell-like pattern typical of a brittle amorphous alloy was seen in Fig. 1.34. The amorphous compacts appear to maintain good ductility without embrittlement. For warm extrusion in the supercooled liquid state and at a small extrusion ratio of 3 or more, it is possible to synthesize bulk amorphous compacts with full tensile strength and ductility. This is the first instance of fabricating amorphous alloy compacts with the same σ_f as that of the cast bulk amorphous alloy or melt-spun amorphous ribbon.

Fig. 1.34. Tensile fracture surface of $Zr_{65}Al_{10}Ni_{10}Cu_{15}$ amorphous compact produced by extraction at T_e=673 K and R=5

Table 1.5. Fundamental characteristics and application fields of bulk amorphous alloys

Fundamental characteristic	Application
	Machinery structural materials
High strength	Optical precision materials
High hardness	Tool materials
High impact fracture energy	Cutting materials
High elastic energy	Corrosion resistant materials
High corrosion resistance	Ornamental materials
High wear resistance	Composite materials
High viscous flowability	Writing appliance materials
Soft magnetism	Sporting goods materials
High magnetostriction	Soft magnetic materials
	High magnetostrictive materials

1.13 Applications and Future Prospects

As described above, the discovery of a number of new alloys with high stability of supercooled liquid to crystallization has induced a drastic decrease by six to eight orders in the critical cooling rate for glass formation and enabled the production of bulk amorphous alloys with thicknesses ranging from several to about 80 mm by various casting processes. This drastic increase in the stability of a supercooled liquid also implies the advent of new basic science and materials. Table 1.5 summarizes the fundamental characteristics and applications of bulk amorphous alloys produced by conventional casting processes. The bulk amorphous alloys exhibit various characteristics, such as high mechanical strength, high elastic energy, high impact fracture energy, high wear resistance, high corrosion resistance, good soft magnetic properties, high frequency permeability, fine and precise viscous deformability, good castability, and high consolidation into bulk forms. By utilizing these characteristics inherent in bulk amorphous alloys, the new alloys are expected to open up new applications in the near future.

References

1. W. Klement, R.H. Willens, and P. Duwez: Nature, **187**, 869 (1960)
2. H. S. Chen: Rep. Prog. Phys., **43**, 353 (1980)
3. Materials Science of Amorphous Metals, ed. by T. Masumoto, (Ohmu Pub., Tokyo 1982)
4. Amorphous Metallic Alloys, ed. by F.E. Luborsky, (Butterworths, London 1983)
5. Rapidly Solidified Alloys, ed. by H.H. Liebermann (Marcel Dekker, Inc., New York 1993)
6. Y.H. Kim, A. Inoue, and T. Masumoto: Mater. Trans., JIM, **31**, 747(1990)
7. H. Chen, Y. He, G.J. Shilet, and S.J. Poon: Scripta Met., **25**, 1421 (1991)
8. Y. Yoshizawa, S. Oguma, and K. Yamauchi: J. Appl. Phys., **64**, 6044 (1988)
9. K. Suzuki, A. Makino, N. Kataoka, A. Inoue, and T. Masumoto: Mater. Trans., JIM, **32**, 93 (1991)
10. J. J. Croat: Appl. Phys. Lett. **37**, 1096 (1980)
11. E. F. Kneller, and R. Hawig: IEEE Trans. Magn., **27**, 3588 (1991)
12. A. Inoue, A. Takeuchi, A. Makino, and T. Masumoto: Mater. Trans., JIM, **36**, 676 (1995)
13. A. Inoue, Y. Tanaka, Y. Miyauchi, and T. Masumoto: Crystallization Behavior of Amorphous Fe-Tb-M (M=Si or Al) Alloys, and High Magnetostriction of their Crystallized Phases, Sci. Rep. Res. Inst. Tohoku Univ., A39,147-153 (1994)
14. A. Yokoyama, H. Komiyama, H. Inoue, T. Masumoto, and H.M. Kimura: J. Catalysis, **68**, 355 (1981)
15. H. S. Chen: Mater. Sci. Eng., **25**, 59 (1976)
16. H. S. Chen: J. Appl. Phys., **49**, 3289 (1978)
17. H. W. Kui, A. L. Greer, and D. Turnbull: Appl. Phys. Lett., **45**, 615 (1984)
18. H. W. Kui, and D. Turnbull: Appl. Phys. Lett., **47**, 796 (1985)
19. A. Inoue, K. Ohtera, K. Kita, and T. Masumoto: Jpn. J. Appl. Phys., **27**, L2248 (1988)
20. A. Inoue, T. Zhang, and T. Masumoto: Mater. Trans., JIM, **30**, 965 (1989)
21. A. Inoue, T. Zhang, and T. Masumoto: Mater. Trans., JIM, **31**, 177 (1990)
22. A. Inoue, T. Zhang, and T. Masumoto: J. Non-Cryst. Solids, **156-158**, 473 (1993)

23. K. Amiya, N. Nishiyama, A. Inoue, and T. Masumoto: Mater. Sci. Eng., **A179/A180**, 692 (1994)
24. A. Peker, and W.L. Johnson: Appl. Phys. Lett., **63**, 2342 (1993)
25. A. Inoue, T. Shibata, and T. Zhang: Mater. Trans., JIM, **36**, 1420 (1995)
26. A. Inoue, N. Nishiyama, and T. Matsuda: Mater. Trans., JIM, **37**, 181 (1996)
27. A. Inoue, Y. Shinohara, and J.S. Gook: Mater. Trans., JIM, **36**, 1427 (1995)
28. Yi He, and R. B. Schwarz: Met. Trans.
29. A. Inoue, T. Zhang, and T. Itoi: Mater. Trans., JIM, **38**, 359 (1997)
30. A. Inoue, H. Koshiba, T. Zhang, and A. Makino: Mater. Trans., JIM, **38**, 577 (1997)
31. A. Inoue, T. Zhang, and A. Takeuchi: Appl. Phys. Lett., **71**, 464 (1997)
32. A. Inoue, H. Koshiba, T. Itoi, and A. Makino: Appl. Phys. Lett., **73**, 744 (1998)
33. A. Inoue, and T. Zhang: Mater. Trans., JIM, **37**, 185 (1996)
34. W. L. Johnson: Mater. Sci. Forum, **225-227**, 35 (1996)
35. A. Inoue, N. Nishiyama, and H. M. Kimura: Mater. Trans., JIM, **38**, 179 (1997)
36. H. A. Davies: "Amorphous Metallic Alloys", ed. F.E. Luborsky (Butterworths, London 1983) p. 14
37. A. Inoue: Mater. Trans., JIM, **36**, 866 (1995)
38. A. Inoue: Mater. Sci. Forum, **179-181**, 691-700 (1995)
39. A. Inoue: Sci. Rep. Res. Inst. Tohoku Univ., **A42**, 1-11 (1996)
40. A. Inoue: Proc. Japan Acad., Ser.B, 19-24 (1997)
41. A. Inoue: Mater. Sci. Eng., **A226-228**, 357 (1997)
42. D. Turnbull: Solid State Phys., **3**, 225 (1956)
43. E. Matsubara, T. Tamura, Y. Waseda, A. Inoue, M. Kohinata, and T. Masumoto: Mater. Trans., JIM, **31**, 228 (1990)
44. E. Matsubara, T. Tamura, Y. Waseda, A. Inoue, T. Zhang, and T. Masumoto: Mater. Trans., JIM, **33**, 873 (1992)
45. E. Matsubara, T. Tamura, Y. Waseda, T. Zhang, A. Inoue, and T. Masumoto: J. Non-Cryst. Solids, **150**, 380 (1992)
46. A. Inoue, H. M. Kimura, T. Negishi, T. Zhang, and A. R. Yavari: Mater. Trans., **39**, 318 (1998)
47. T. Masumoto, H. M. Kimura, A. Inoue, and Y. Waseda: Mater. Sci. Eng., **23**, 141 (1976)
48. A. Inoue, D. Kawase, A. P. Tsai, T. Zhang, and T. Masumoto: Mater. Sci. Eng., **A178**, 225 (1994)
49. C. V. Thomson, A. L. Greer, and D. Spaepen: Acta Met. Mater., **31**, 1883 (1983)
50. Z. Stmad: "Glass-Ceramic Materials" (Elsevier, Amsterdam 1986) p. 114
51. N. Nishiyama, and A. Inoue: Mater. Trans., JIM, **37**, 1531 (1996)
52. N. Nishiyama, and A. Inoue: Mater. Trans., JIM, **38**, 464 (1997)
53. D. R. Uhlmann: J. Non-Cryst. Solids, **7**, 337 (1972)
54. H. A. Davies: J. Non-Cryst. Solids, **17**, 266 (1975)
55. D. R. Uhlmann: Materials Science Research, Vol.4, Plenum, New York (1969), p. 172.
56. N. Nishiyama, Doctor Thesis, Tohoku Univ., (1997)
57. J. W. Christian: The Theory of Phase Transformation in Metals and Alloys, Pergamon Press, Oxford (1965), p. 377

58. A. Inoue, T. Zhang, K. Ohba, and T. Shibata: Mater. Trans., JIM, **36**, 876 (1995)
59. A. Inoue, Y. Yokoyama, Y. Shinohara, and T. Masumoto: Mater. Trans., JIM, **35**, 923 (1994)
60. A. Inoue, Y. Shinohara, and Y. Yokoyama: Mater. Trans., JIM, **36**, 1276 (1995)
61. A. Inoue, and T. Zhang: Metals, **3**, 47 (1994)
62. W. Kurz, and D. J. Fisher: "Fundamentals of Solidification" (Trans Tech Publications, Zürich 1989)
63. M. C. Flemings: "Solidification Processing" (McGraw-Hill, New York 1974)
64. R. Trivedi, and W. Kurz: Acta Met., **34**, 1663 (1986)
65. A. Inoue, T. Zhang, and T. Masumoto: J. Non-Cryst. Solids, **150**, 396 (1992)
66. H. S. Chen: J. Non-Cryst. Solids, **46**, 289 (1981)
67. H. S. Chen, A. Inoue, and T. Masumoto: J. Mater. Sci., **20**, 4057 (1985)
68. A. Inoue, T. Masumoto, and H. S. Chen: J. Mater. Sci., **20**, 2417 (1985)
69. H. S. Chen, and K. A. Jackson, Metallic Glasses, ASM, Ohio, (1978), p. 75
70. T. Zhang, A. Inoue, and T. Masumoto: Mater. Trans., JIM, **32**, 1005 (1991)
71. O. Haruyama, H. M. Kimura, and A. Inoue: Mater. Trans., JIM, **37**, 1741 (1996)
72. H. Okumura, A. Inoue, and T. Masumoto: Mater. Trans., JIM, **32**, 599 (1991)
73. H. M. Kimura, M. Kishida, T. Kaneko, A. Inoue, and T. Masumoto: Mater. Trans., JIM, **36**, 890 (1995)
74. H. M. Kimura, A. Inoue, N. Nishiyama, K. Sasamori, O. Haruyama, and T. Masumoto: Sci. Rep. Res. Inst. Tohoku Univ., **A43**, 101-106 (1997)
75. T. Masumoto, and H. M. Kimura: Sci. Rep. Res. Inst. Tohoku Univ., **A25**, 216-231 (1975)
76. J. H. Mooij: Phys. Status Solidi, **A17**, 521 (1973)
77. Ch. Ed. Guillaume: Comptes Rendus de l'Academie des Sciences, Paris, 171(1920), p.1039
78. A. Inoue, T. Zhang, and T. Masumoto: Mater. Trans., JIM, **36**, 391 (1995)
79. A. Inoue, and T. Zhang: Mater. Trans., JIM, **37**, 1726 (1996)
80. K. Ohtera, Doctor Thesis, Tohoku Univ., (1996)
81. A. Inoue: Bulletin Japan Inst. Metals, **36**, 926 (1997)
82. L. E. Nielsen: "Mechanical Properties of Polymers" (Chap. 7, Reinhold, New York 1962)
83. H. Okumura, A. Inoue, and T. Masumoto: Acta Metall. Mater., **41**, 915 (1993)
84. H. Okumura, H.S. Chen, A. Inoue, and T. Masumoto: J. Non-Cryst. Solids, **142**, 165 (1992)
85. H. E. Kissinger: Analyt. Chem., **29**, 1702 (1957)
86. H. Okumura, A. Inoue, and T. Masumoto: Mater. Trans., JIM, **32**, 599 (1991)
87. S. Iwayanagi, and T. Hideshima: J. Phys. Soc. Japan, **8**, 365 (1953)
88. H. Okumura, A. Inoue, and T. Masumoto: Sci. Rep. Res. Inst. Tohoku Univ., **A36**, 239-260 (1991/92)
89. H. Okumura, A. Inoue, and T. Masumoto: Jpn. J. Appl. Phys., **31**, 3403 (1992)
90. A. Inoue, A. Takeuchi, T. Zhang, A. Murakami, and A. Makino: IEEE Trans. Magn., **32**, 4866 (1996)
91. A. Inoue, and A. Katsuya: Mater. Trans., JIM, **37**, 1332 (1996)

92. A. Inoue, H. Koshiba, T. Zhang, and A. Makino, J. Appl. Phys., **83**, 1967 (1998)
93. A. Inoue, A. Murakami, T. Zhang, and A. Takeuchi: Mater. Trans., JIM, **38**, 189 (1997)
94. A. Inoue, and R. E. Park: Mater. Trans., JIM, **37**, 1715 (1996)
95. M. Kikuchi, H. Fujimori, T. Obi, and T. Masumoto: Jpn. J. Appl. Phys., **14**, 1077 (1975)
96. C. H. Smith, "Rapidly Solidified Alloys", ed. H.H. Liebermann (Marcel Dekker, New York 1993) p. 617
97. "Materials Science of Amorphous Alloys", ed. T. Masumoto (Ohmu, Tokyo 1982) p. 97
98. "Metals Databook", ed. Japan Inst. Metals (Maruzen, Tokyo 1981), p. 121
99. A. Inoue, and Y. Saotome: Metals, **3**,51 (1993)
100. A. Inoue, Y. Kawamura, and Y. Saotome: Mater. Sci. Forum, **233-234**, 147 (1997)
101. A. Inoue, and T. Zhang: Mater. Sci. Forum, **243-245**, 197-206 (1997)
102. A. Inoue, A. Takeuchi and T. Zhang, Proc. Int. Conf. and Exhib., "Micro Mat '97", 103 (1998)
103. A. Inoue, Y. Kawamura, T. Shibata, and K. Sasamori: Mater. Trans., JIM, **37**, 1337 (1996)
104. Y. Kawamura, T. Shibata, A. Inoue, and T. Masumoto: Appl. Phys. Lett., **69**, 1208 (1996)
105. N. Nishiyama, and A. Inoue, Mater. Trans., JIM, **40**, 64 (1999)
106. T. Zhang, A.P. Tsai, A. Inoue, and T. Masumoto: Boundary, 7, No.9, p. 39-43 (1991)
107. S. Hata, Y. Saotome, and A. Inoue
108. "Amorphous Metallic Alloys", ed. F.E. Luborsky, (Butterworths, London 1983)
109. H. Kato, Y. Kawamura, and A. Inoue: Mater. Trans., JIM, **37**, 70 (1996)
110. Y. Kawamura, H. Kato, A. Inoue, and T. Masumoto: Appl. Phys. Lett., **69**, 1208 (1996)
111. Y. Kawamura, H. Kato, A. Inoue, and T. Masumoto: Int. J. Powder Metallurgy, **33**, 50(1997)
112. Y. Kawamura, A. Inoue, K. Sasamori, and T. Masumoto: Mater. Sci. Eng., **A181/A182**, 1174 (1994)
113. R. J. Wilcox, and P. W. Whitton: J. Inst. Met., **87**, 289 (1958-59)

2 Stress Relaxation and Diffusion in Zr-Based Metallic Glasses Having Wide Supercooled Liquid Regions

Y. Kawamura[1], T. Shibata[1], A. Inoue[1], T. Masumoto[2], K. Nonaka[3], H. Nakajima[4], and T. Zhang[1]

[1]Institute for Materials Research, Tohoku University, Aoba-ku, Sendai 980-8577, Japan
[2]The Research Institute for Electric and Magnetic Materials, Taihaku-ku, Sendai 982-0807, Japan
[3]Department of Materials Science and Technology, Iwate University, Morioka 020-8551, Japan
[4]The Institute of Scientific and Industrial Research, Osaka University, Ibaraki, Osaka 567-0047, Japan

Summary. We investigated the stress relaxation behavior and diffusion of Zr-Al-Ni-Cu metallic glasses having a wide supercooled liquid region. The stress relaxation was more pronounced after yielding and its relaxation rate increased with temperature. The stress relaxation was associated with the stress overshoot that is a transient stress-strain phenomenon. The stress overshoot appeared again after the stress relaxation in the course of stretch, and increased with an increase in stress relaxation fraction. The glassy structure appears to change into a state with higher atomic mobility by yielding, and to return its initial structure by the stress relaxation. The temperature dependence of the diffusivity in the supercooled liquid phase above the glass transition temperature was significantly different from that in the amorphous phase. The activation energy for diffusion in the supercooled liquid phase was much larger than that in amorphous phase below the glass transition temperature.

2.1 Introduction

Recently, metallic glasses having excellent glass-forming ability and wide supercooled liquid region more than 60 K were discovered [1–7]. These metallic glasses also have useful mechanical properties such as high strength, high impact fracture energy, high toughness and high-strain-rate superplasticity [8–16]. In order to use the metallic glasses as structural materials, deep understanding of the deformation behavior is required. Moreover, various thermal properties such as deformation behavior, structural relaxation, glass transition and crystallization are known to be controlled by atomic diffusion processes. Therefore, knowledge of diffusion is important for the understanding of various processes occurring in the materials. However, there has been quite few investigations on self- and impurity diffusion in a wide supercooled liquid region of amorphous alloys: Au diffusion in $Pd_{77.5}Cu_6Si_{16.5}$ metallic glass [17], and Be diffusion in metallic glass [18]. The accumulation of reliable experimental data is important to understand the mechanism of diffusion in the metallic glasses. In this study, we will report the stress relaxation behavior of $Zr_{65}Al_{10}Ni_{10}Cu_{15}$ metallic glass and the result of diffusion measurements of Ni in amorphous phase and supercooled liquid phase of $Zr_{55}Al_{10}Ni_{10}Cu_{25}$ metallic glass.

2.2 Experiments

Alloy ingots were prepared by arc-melting mixtures of pure elements on a water cooled copper hearth under an argon gas atmosphere using Ti as an oxygen getter. $Zr_{65}Al_{10}Ni_{10}Cu_{15}$ metallic glass ribbons with a cross section of about 1.00×0.02 mm^2 were produced by a single-roller melt-spinning method under an argon gas atmosphere for stress relaxation tests. $Zr_{55}Al_{10}Ni_{10}Cu_{25}$ bulk metallic glasses were produced by a copper-mold casting method for measurments of diffusivity. The formation of a single glassy phase was confirmed by X-ray diffractometry, transmission electron microscopy (TEM) and compositional imaging an electron probe micro analyzer (EPMA). Thermal properties were measured by differential scanning calorimetry (DSC) and differential thermal analysis (DTA).

Specimens for stress relaxation tests were prepared by gluing the metallic glass ribbon on the ceramic holders with a ceramic cement (Sauereisen Cement), and the tensile tests were conducted using an Instron-type tensile test apparatus. The gauge length was 10 mm. A thermocouple was placed closely to the sample. The experiments were started after allowing the sample to equilibrate for 200 s. The pull rods used in the tensile test apparatus were made from a maraging steel. The average thermal expansion coefficient of the steel is about 13×10^{-6} K^{-1} in the range from 20 to 800 K. Since the length of the tubular furnace was 400 mm, the pull rods were estimated to extend by 5.2×10^{-3} mm/K. The fluctuation of temperature during the relaxation tests was ±1.0 K, resulting in a fluctuation of less than ±0.05 % in the strain of the samples.

Rectangular shaped samples with $10\times8\times2$ mm^3 for the measurements of diffusivity were cut from the casted bulk metallic glasses. The sample surfaces were prepared by standard metallographical procedures on lapping tape and diamond paste down to 1 µm. Then, the samples were annealed in vacuum for 300s at 685K in the supercooled liquid region for the structural relaxation. The radioisotope ^{63}Ni was purchased from E.I. Du Pont de Nemours & Co. Inc. in the form of nickel chloride in 0.5M hydrochloric acid. The each solution was diluted with dimethyl sulfoxide, which was used for electroplating. The electroplating was carried out to deposit ^{63}Ni isotopes onto the mirror polished surface of the disc specimens at a current density of 1.9 Am^{-2} for 600 s; a Pt loop anode was just touched to the top of about ten drops of the solution placed on the specimen surface. For annealing below T_g, the specimen was wrapped with Ta foils and was sealed in a quartz tube in vacuum of 10^{-3} to 10^{-4} Pa. The diffusion anneals were carried out at given temperatures controlled to within ± 1K. For shorter anneals above T_g, the sample was annealed in a quartz tube evacuated by the Turbomolecular pump until 10^{-5} Pa and heated by an infrared furnace. The heating rate was controlled at 0.67 K/s. Correction for the diffusion-annealing time was made by taking into account the heating and cooling times. After diffusion anneals, the specimens were sectioned by the ion-beam sputter-sectioning apparatus. The sputtered material was deposited onto Mylar foil. Serial sectioning of the whole penetration profiles was performed by moving the collector foil step by step. Typically 20 to 30 sections were collected. The specimen films were dissolved in 0.8 ml of the solution of 23% hydrofluoric acid, 38% nitric acid and 39% distilled water. The Aquasol-II scintillator of 12 ml was added to each vial for the specimen. Then, the β-ray activity from ^{63}Ni was measured by a liquid scintillation counter (Aloka LSC-5010).

2.3 Results and Discussion

2.3.1 Stress Relaxation in $Zr_{65}Al_{10}Ni_{10}Cu_{15}$ Metallic Glass

Figure 2.1 shows DSC curve of the $Zr_{65}Al_{10}Ni_{10}Cu_{15}$ metallic glass at a scanning rate of 0.67 K/s. The glass transition temperature (T_g), crystallization temperature (T_x) and supercooled liquid region $(\Delta T_x = T_x - T_g)$ of the

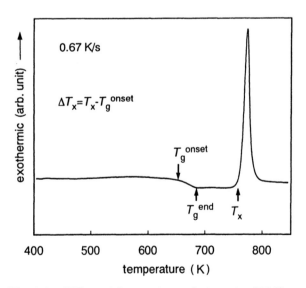

Fig. 2.1. Differential scanning calorimetric (DSC) curve of $Zr_{65}Al_{10}Ni_{10}Cu_{15}$ metallic glass at a heating rate of 0.67 K/s. The T_g^{onset}, T_g^{end} and T_x represent the onset and end temperatures of the glass transition and the crystallization temperature, respectively

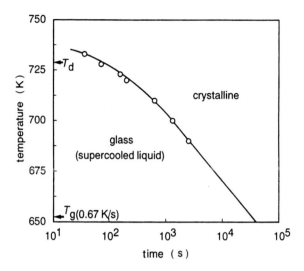

Fig. 2.2. Time-temperature-transformation $(T-T-T)$ diagram for the onset of crystallization in $Zr_{65}Al_{10}Ni_{10}Cu_{15}$ metallic glass

metallic glass were 652 K, 757 K and 105 K, respectively. The melting temperature (T_m) measured by DTA was 1121 K. The temperatures of T_g and T_x correspond to $0.58T_m$ and $0.68T_m$, respectively. Figure 2.2 shows the time-temperature-transformation $(T-T-T)$ diagram for the onset of crystallization. The incubation time for retaining the glassy phase without decomposition at the glass transition temperature was about 3×10^4 s.

Figure 2.3 shows the tensile stress-strain curves of a $Zr_{65}Al_{10}Ni_{10}Cu_{15}$ metallic glass ribbon recorded at a strain rate of 5.0×10^{-4} s^{-1} and at 573,

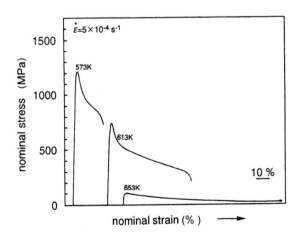

Fig. 2.3. Tensile stress-strain curves at temperatures of 573, 613 and 653 K and at a strain rate of 5.0×10^{-4} s^{-1} in $Zr_{65}Al_{10}Ni_{10}Cu_{15}$ glassy ribbon

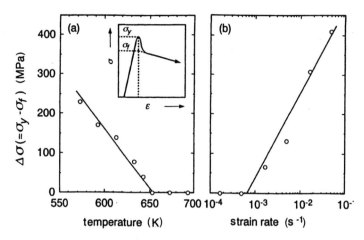

Fig. 2.4. Changes in the stress overshoot, which is the difference between the yield stress and the stable flow stress as shown in the inset, as a function of testing temperature at a strain rate of 5×10^{-4} s^{-1} (a), and as a function of strain rate at 653 K (b) in $Zr_{65}Al_{10}Ni_{10}Cu_{15}$ glassy ribbon

613 and 653 K. A stress overshoot was recognized in the stress-strain curves at 573 and 613 K. The stress overshoot was a transient stress-strain phenomenon in the homogeneous deformation mode. After the stress overshoot, the flow reached a steady state, namely, the plateau of stress in the stress-strain curves. Figure 2.4 summarizes the strain-rate and temperature dependence of the stress overshoot. At a strain rate of 5×10^{-4} s^{-1}, the stress overshoot decreased with temperature and approached zero at temperatures higher than 653 K where the metallic glass is in a supercooled liquid state. On the other hand, at a temperature of 653 K the stress overshoot began to appear at strain rates higher than 1×10^{-3} s^{-1} and increased with strain rate. The subsequent stress relaxation tests were carried out below and above the yield strain ε_y where the stress exhibited a peak and yielding occurred. Here, we defined the peak stress as a yield stress (σ_y).

Figure 2.5 shows the stress relaxation curves at 573, 613 and 653 K, where the samples were first stretched at a constant strain rate of 5×10^{-4} s^{-1} and then held at a strain of $\varepsilon_y + 10$ %. The stress values during the relaxation were normalized with the stresses (σ_0) at the time when the cross head was stopped. As shown in Fig. 2.5, the stress relaxation progressed at higher temperatures. Figure 2.6 exemplifies the stress-relaxation curves at 573 K and at several strains. Below the yield point the stress relaxation was measured at strains corresponding to $0.25\sigma_0$, $0.50\sigma_0$, $0.75\sigma_0$ and $1.00\sigma_0$, and above the yield point at $\varepsilon_y + 10$ % and $\varepsilon_y + 20$ % strains, as schematically illustrated in the inset of Fig. 2.6. The samples were first loaded at a constant strain rate of 5.0×10^{-4} s^{-1} and then kept at the above-mentioned strains. The stress measured during the relaxation was normalized with the initial stress (σ_0).

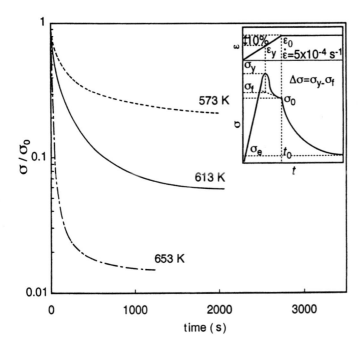

Fig. 2.5. Stress relaxation curves at a strain of $\varepsilon_y + 10\%$ and at temperatures of 573, 613 and 653 K in $Zr_{65}Al_{10}Ni_{10}Cu_{15}$ glassy ribbon

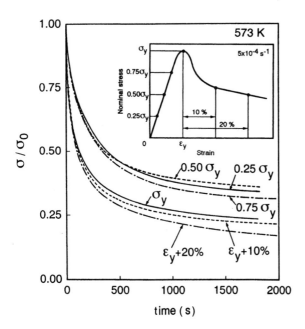

Fig. 2.6. Stress relaxation curves at several strains and at 573 K in $Zr_{65}Al_{10}Ni_{10}Cu_{15}$ glassy ribbon

One should not overlook that the feature of the stress relaxation changed after the yielding. The yielding seems to promote the stress relaxation and lead to a decrease of the normalized stress level.

Figure 2.7 shows the $\log(\sigma/\sigma_0)$ versus $\log t$ plots of the stress relaxation at several strain levels at temperatures of 573, 613 and 653 K. The stress relaxation behavior is usually divided into two stages. The first stage ($t < \lambda_c$) extends to the relaxation time (λ_c) required for the stress to decrease to one neper, i.e. σ_0/e. The stress relaxation in the first stage may be approximated by the Kohlrausch-Williams-Watts (KWW) equation [19]

$$\sigma/\sigma_0 = \exp[-(t/\lambda_c)^\beta], \tag{2.1}$$

where β is a fitting parameter, corresponding to a stress relaxation rate. The second stage ($t > \lambda_c$) usually exhibits a linear behavior in the $\log(\sigma/\sigma_0)$ versus $\log t$ plots. The stress relaxation in the second stage may be approximated by [19]

$$\sigma/\sigma_0 = (1/e)(t/\lambda_c)^\beta). \tag{2.2}$$

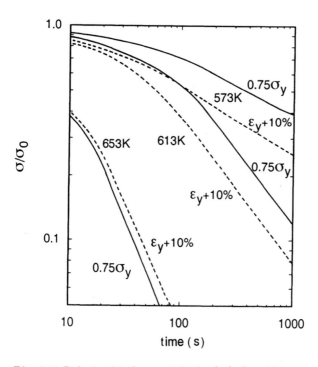

Fig. 2.7. Relationship between the $\log(\sigma/\sigma_0)$ and $\log t$ at several strains above and below the yield point at temperatures of 573, 613 and 653 K in a $Zr_{65}Al_{10}Ni_{10}Cu_{15}$ glassy ribbon

As shown in Fig. 2.7, the value of β, namely, the slope in the second stage was unchanged by yielding and increased with temperature. The β may be approximated by the Vogel-type equation [19]

$$\beta = A(T - T_0)/T^2, \tag{2.3}$$

where A is a fitting parameter and T_0 is a reference temperature. Figure 2.8 shows the β versus $(T - T_0)/T^2$ plots for the stress relaxation at $0.75\sigma_y$ and $\varepsilon_y+10\%$, where T_0 was selected as 552 K (T_g-100 K). The plots exhibited a linear relationship with a similar value of A which is 5500 ± 900 before the yielding and 4900 ± 700 after the yielding, showing that the stress relaxation rate β is independent of the strain level in the stress relaxation. The relaxation time, however, depends on the strain level. Stress relaxation is a thermally activated process. Hence its activation energy can be calculated by the Arrhenius-type equation

$$\lambda_c = \lambda_0 \exp(-H/RT), \tag{2.4}$$

where λ_0 and R are a fitting parameter and a gas constant, respectively, and H is the activation energy of the stress relaxation. Figure 2.9 shows

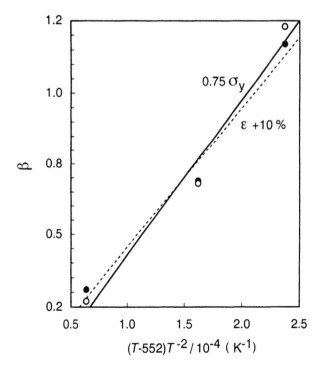

Fig. 2.8. Relationship between the β and $(T - T_0)/T^2$ for the stress relaxation at strains having $0.75\sigma_0$ and $\varepsilon_y+10\%$ in $Zr_{65}Al_{10}Ni_{10}Cu_{15}$ glassy ribbon

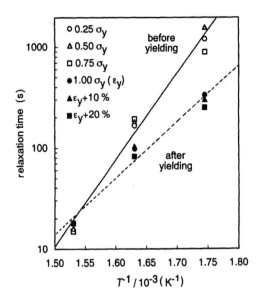

Fig. 2.9. Relationship between the log λ_c and $1/T$ for the stress relaxation at strains having $0.75\sigma_0$ and $\varepsilon_y+10\%$ in $Zr_{65}Al_{10}Ni_{10}Cu_{15}$ glassy ribbon

the log λ_c versus $1/T$ plots for the stress relaxations at several strain levels. The activation energies were estimated to be about 108±8 kJ/mol after the yielding, and about 165±10 kJ/mol before the yielding. Accordingly, it was found that the yielding reduced the activation energy by about 60 kJ/mol. This means that the structure in the yielded metallic glass changed to a state which has a higher atomic mobility than that in the unyielded metallic glass. On the other hand, the relaxation time after the yielding is independent of the strain levels. Moreover, the stress in the stress-strain curves is transformed into a steady flow state. After yielding the metallic glass, therefore, seems to deform plastically while having the same structure which is identical at each temperature and strain rate. The change in the structure by yielding appears to contribute to the stress overshoot. The metallic glasses which reached the steady flow state exhibit again a stress overshoot when restressed after stress relaxation. The stress-relaxation fraction dependence of the stress overshoot is shown in Fig. 2.10. The stress overshoot increased with an increase in the stress relaxation fraction. This phenomenon probably means that during stress relaxation the structure with a higher atomic mobility turns back to the structure which is similar to that before the yielding.

The deformation of metallic glasses, which depends on temperature and strain rate, is divided into two modes, namely, inhomogeneous mode and homogeneous mode. The inhomogeneous deformation, which is observed at lower temperatures or higher strain rates, is localized in discrete and thin shear bands, resulting from its nonhardenable nature. In the homogeneous

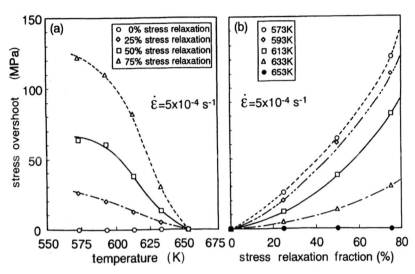

Fig. 2.10a,b. Stress-relaxation fraction dependence of the stress overshoot which is appeared after the stress-relaxation in the course of the stretch at a strain rate of 5.0×10^{-4} s^{-1}

deformation which is observed at higher temperatures or lower strain rates, each volume element of materials contributes to strain, resulting in a uniform deformation for uniformly stressed specimen. It has been reported that there are changes in the mechanical properties of the shear bands which are formed by inhomogeneous deformation [20,21]. When the samples previously deformed in the inhomogeneous deformation mode are restressed, the deformation occurs at the same shear bands formed on the first inhomogeneous deformation. This phenomenon observed in the inhomogeneous deformation mode seems to agree with the present results that were obtained in the homogeneous deformation mode.

For further deep understanding of the stress relaxation behavior, the characterization of the structure of the plastically deformed and undeformed metallic glasses using small-angle X-ray scattering (SAXS) and neutron scattering will be subjected.

2.3.2 Diffusion in $Zr_{55}Al_{10}Ni_{10}Cu_{25}$ Metallic Glass

Figure 2.11 shows the X-ray diffraction pattern taken from the polished surface of the sample. The diffraction pattern consists only of halo rings, indicating a glassy phase. In order to confirm the absence of a crystalline phase over the whole sample, compositional image by EPMA taken from the middle region is shown in Fig. 2.12. No contrast reveals that no precipitation of a crystalline phase is seen over the whole micrograph. Furthermore, no cavities and holes were observed. Figure 2.13 shows the DSC curve of $Zr_{55}Al_{10}Ni_{10}Cu_{25}$

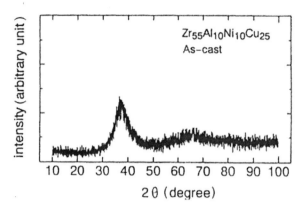

Fig. 2.11. X-ray diffraction pattern taken from $Zr_{55}Al_{10}Ni_{10}Cu_{25}$ metallic glass

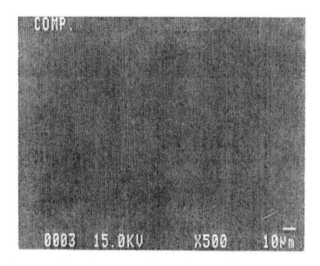

Fig. 2.12. Compositional image of $Zr_{55}Al_{10}Ni_{10}Cu_{25}$ metallic glass

metallic glass sample. The result shows a distinct endothermic reaction due to the glass transition, followed by a supercooled liquid region and then an exothermic reaction due to crystallization. The glass transition and the crystallization temperatures were determined to be 683 K and 769 K, respectively. Thus, the interval of the supercooled liquid region is 86 K. Figure 2.14 shows the time-temperature-transformation $(T-T-T)$ diagram. The $T-T-T$ diagram was determined by DSC, which was used to know the stability of the supercooled liquid region. In order to measure the diffusion in single phase of the supercooled liquid region, the diffusion was carried out for annealing times shorter than the onset of the first stage of the crystallization (see in Fig. 2.14: $t < t_{xc}$).

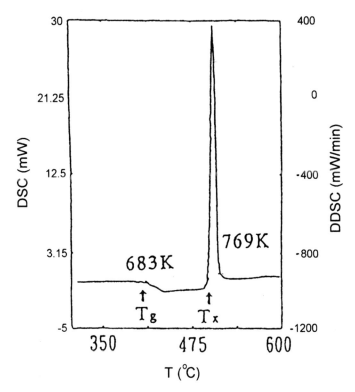

Fig. 2.13. DSC curve of $Zr_{55}Al_{10}Ni_{10}Cu_{25}$ metallic glass. The heating rate is 40 $Kmin^{-1}$

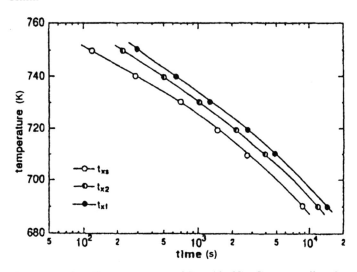

Fig. 2.14. $T - T - T$ diagram of $Zr_{55}Al_{10}Ni_{10}Cu_{25}$ metallic glass

The diffusion condition is equivalent to an infinitely thin source diffusing into a semi-infinite cylinder. The diffusion-penetration profiles confirm well to the thin film solution of the diffusion equation,

$$C(x,t) = M(\pi(Dt)^{-1/2}\exp(-x^2/4Dt), \tag{2.5}$$

where $C(x,t)$ is the tracer concentration at a depth x after a diffusion interval t, D is the tracer diffusivity, and M is the initial amount of a tracer at the surface. Figures 2.15 and 2.16 show the penetration profiles for the diffusion of ^{63}Ni in the amorphous phase and the supercooled liquid phase of $Zr_{55}Al_{10}Ni_{10}Cu_{25}$ metallic glass, respectively. All diffusion profiles were Gaussian without serious surface hold-up or noticeable non-Gaussian tails. These penetration profiles were analyzed by using a least square fitting to (2.5). The temperature dependence of the diffusivities of Ni in $Zr_{55}Al_{10}Ni_{10}Cu_{25}$ metallic glass is shown in Fig. 2.17. The temperature dependence drastically change at the glass transition temperature. The diffusivity in the supercooled liquid phase is much higher than that extrapolated from the data in the amorphous phase. Fitting the data to the Arrhenius equation, the temperature dependence can be expressed as

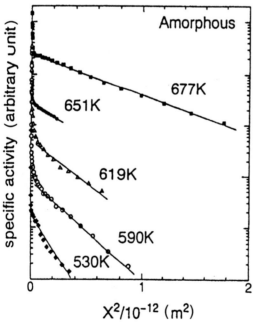

Fig. 2.15. Penetration profiles for diffusion of ^{63}Ni in the amorphous phase of $Zr_{55}Al_{10}Ni_{10}Cu_{25}$ metallic glass

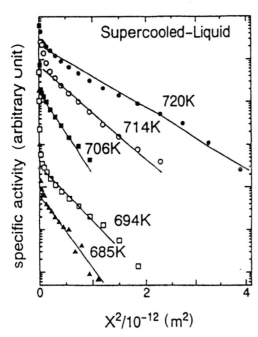

Fig. 2.16. Penetration profiles for diffusion ^{63}Ni in the supercooled liquid phase of $Zr_{55}Al_{10}Ni_{10}Cu_{25}$ metallic glass

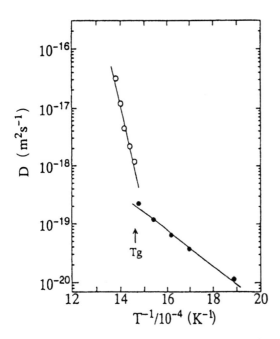

Fig. 2.17. Temperature dependence of the diffusivities of Ni in $Zr_{55}Al_{10}Ni_{10}Cu_{25}$ metallic glass

in amorphous phase

$$D\ \text{amorph} = 6.8 \times 10^{-15} \exp(-59 \pm 2\ \text{kJmol}^{-1}/RT)\ \text{m}^2\text{s}^{-1}$$

in supercooled liquid phase

$$D\ \text{supercool} = 4.5 \times 10^9 \exp(-363 \pm 32\ \text{kJmol}^{-1}/RT)\ \text{m}^2\text{s}^{-1}.$$

Figure 2.18 shows the diffusivities of Ni versus inverse of temperature in $Zr_{55}Al_{10}Ni_{10}Cu_{25}$ metallic glass, together with previous data on Be diffusion in $Zr_{41.2}Ti_{13.8}Ni_{10}Cu_{12.5}Be_{22.5}$ [18] and Au diffusion in $Pd_{77.5}Cu_6Si_{16.5}$ metallic glass [17]. The temperature dependence in the present work is similar to that by Geyer et al. [18]. Geyer et al. [18] explained that below the glass transition temperature a single atomic jump diffusion occurs, while Be atoms diffuse by single atomic jumps in a slowly changing configuration of neighbouring atoms above the glass transition. This change of the mechanism is due to the change in configurational entropy through the glass transition. However, the contribution of the saddle point configurational entropy and the migration enthalpy should be taken into account. Further investigations will be published elsewhere to elucidate the diffusion mechanism in the supercooled liquid region and amorphous region [20].

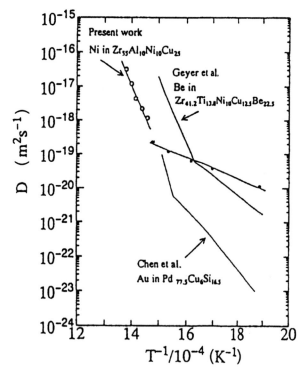

Fig. 2.18. Temperature dependence of diffusion coefficients in various metallic glass

2.4 Conclusions

We have investigated the stress relaxation behavior and diffusion of Zr-Al-Ni-Cu metallic glasses. The results obtained are summarized as follows:

(1) The stress relaxation was accelerated with temperature. Moreover, the yielding promoted the stress relaxation. The activation energy of the relaxation decreased by 60 kJ/mol and corresponded to 108±8 kJ/mol for the yielded samples. The yielding seems to change the structure to a state having high activity in atomic mobility. The stress relaxation was associated with the stress overshoot. The stress overshoot which was appeared again after the stress relaxation in the course of stretch increased with an increase in stress relaxation fraction. The structure with higher atomic mobility which is obtained by yielding seems to turn back to the structure which is similar to that before yielding.

(2) The temperature dependence of the diffusivities of Ni in $Zr_{55}Al_{10}Ni_{10}Cu_{25}$ metallic glass drastically change at the glass transition temperature. The diffusivity in the supercooled liquid phase is much higher than that extrapolated from the data in the amorphous phase. The contribution of the saddle point configurational entropy and the migration enthalpy should be taken into account.

References

1. A. Inoue, T. Zhang and T. Masumoto: Mater. Trans. JIM, **31**, 17 (1990)
2. A.Inoue, M.Kohinata, K.Ohtera, A.P.Tsai and T.Masumoto: Mater. Trans., JIM, **30**, 378 (1989)
3. A.Inoue, T.Nakamura, T.Sugita, T.Zhang and T.Masumoto: Mater. Trans., JIM, **34**, 351 (1993)
4. A. Inoue and J. S. Gook: Mater. Trans. JIM, **36**, 1180 (1995)
5. A. Peker and W. L. Johnson: Appl. Phys. Lett., **63**, 2342 (1993)
6. A. J. Drehman, A. L. Greer and D. Turnbull: Appl. Phys. Lett., **4**, 715 (1982)
7. H. S. Chen, Acta. Metall., **22**, 1505 (1974)
8. T. Masumoto and R. Maddin: Acta Metall., **19**, 725 (1971)
9. R. Maddin and T. Masumoto: Mater. Sci. Eng., **9**, 153 (1972)
10. C. A. Pampillo: J. Mater. Sci., **10**, 1194 (1975)
11. F. Spaepen and A. I. Taub, Amorphous Metallic Alloys , edited by F. E. Luborsky,(Butterworths, London,1983), p. 231
12. J. C. M. Li, Rapidly Solidified Alloys , edited by H. H. Liebermann, (Marcel Dekker,Inc, New York, 1993), p. 379
13. Y. Kawamura, T. Shibata, A. Inoue and T. Masumoto: Appl. Phys. Lett., **69**, 1208 (1996)
14. Y. Kawamura, T. Shibata, A. Inoue and T. Masumoto: Scripta Metall. Mater., **37**, 431 (1997)
15. Y. Kawamura, T. Shibata, A. Inoue and T. Masumoto: Appl. Phys. Lett., **71**, 779 (1997)

16. Y. Kawamura, T. Shibata, A. Inoue and T. Masumoto: Acta Mater., **37**, 253 (1998)
17. H. S. Chen, L. C. Kimerling, J. M. Poate and W. L. Brown: Appl. Phys. Lett., **32**, 461 (1978)
18. U. Gyer, S. Schneider, W. L. Johnson, Y. Qui, T. A. Tombrello and M.-P. Macht: Phys. Rev. Lett., **75**, 2364 (1995)
19. S. Matsuoka, *Relaxation Phenomenon in Polymers*, (Oxford Universit Press, New York,1992)
20. H. Nakajima, K. Nonaka, T. Kojima, T. Zhang and A. Inoue, to be published

3 The Anomalous Behavior of Electrical Resistance for Some Metallic Glasses Examined in Several Gas Atmospheres or in a Vacuum

O. Haruyama[1], H. Kimura[2], N. Nishiyama[3], T. Aoki[4], and A. Inoue[2]

[1]Department of Physics, Faculty of Science and Technology, Science University of Tokyo, Noda 278-8510, Japan
[2]Institute for Materials Research, Tohoku University, Aoba-ku, Sendai 980-8577, Japan
[3]Inoue Superliquid Glass Project, ERATO, Japan Science and Technology Corporation, Sendai 980-0807, Japan
[4]Hiranuma Industry Co., Ltd., Mito 310-00, Japan

Summary. The thermal stability in the supercooled liquid region was examined for $Pd_{82}Si_{18}$, $Pd_{76}Cu_6Si_{18}$, $Pd_{40}Ni_{10}Cu_{30}P_{20}$ and $Zr_{60}Al_{15}Ni_{25}$ glasses by means of mainly in-situ electrical resistance measurement carried out under various atmospheres, such as Ar, He, H_2 and vacuum. A clear variation was found out in the slope of electrical resistance curve after glass transition for all glasses. The glass transition and crystallization temperatures corresponded to those obtained by differential scanning calorimetry. No crystallites were detected in a $Zr_{60}Al_{15}Ni_{25}$, $Pd_{76}Cu_6Si_{18}$ and $Pd_{40}Ni_{10}Cu_{30}P_{20}$ glasses heated to the supercooled liquid region at least within X-ray diffraction and transmission electron microscopy. Some anomalous behaviors of electrical resistance were observed around room temperature for Pd-Si based glasses and in the super-cooled liquid region for a $Pd_{40}Ni_{10}Cu_{30}P_{20}$ glass, possibly resulting from hydrogen absorption and desorption. On the other hand, the behavior of electrical resistance for a $Zr_{60}Al_{15}Ni_{25}$ glass was strongly dependent on the surface state of the sample, containing the oxidation. The change in electrical resistance after glass transition was also examined in detail and explained by Faber-Zimann theory for $Pd_{76}Cu_6Si_{18}$ and $Pd_{40}Ni_{10}Cu_{30}P_{20}$ glasses.

3.1 Introduction

Since the metallic glasses with a wide supercooled liquid region, such as a Zr-Ni-Al [1], Zr-Ti-Cu-Ni-Be [2], Mg-Ln-(Ni, Cu) [3], La-Ni-Al [4], Pd-Ni-Cu-P [5] and so on have been found out, a lot of studies have started and are in progress now. The features of these glasses are to exhibit clearly the glass transition phenomenon and have the distinct supercooled liquid region. Although it was important to check whether or not the stability of the supercooled liquid state was really established for these glasses, there were a few reports about it. If the stability of these glass is confirmed to be so high in the supercooled liquid region, then we can examine sufficiently properties peculiar to the supercooled liquid state and glass transition phenomenon itself. It is already established that the electrical resistance or resistivity is one of physical quantities sensitive to the phase transition. The change in the temperature dependence of the electrical resistivity associated with glass transition has been reported for some Pd-Si based glasses [6]. Then the variation observed after glass transition has been attributed to the formation of tiny crystallites which acted as a new scattering center for conduction electrons. We have reported previously that some Zr-based [7] and Pd-Si(B) based [8] glasses showed the curious behavior of electrical resistance curve in the supercooled liquid region, although the corresponding DSC (differential scanning calorimeter) thermogram did not show any exothermic and endothermic peaks in this region. Concerning a Zr-based glass, it was confirmed that the behavior of electrical resistance has been influenced strongly by the oxidation of the sample in the supercooled liquid region. Pd-Si glasses containing B atom have shown also a less stable behavior in the supercooled liquid region, probably attributed to the precipitation of nano-scaled crystallite. In the present experiment, under the prediction that the stability of some glasses were reduced by oxidation of the sample during the measurement, the experiment was performed in H_2 atmosphere and vacuum. It has remained unknown whether or not the change in the electrical resistance through glass transition is caused by the inherent variation of electron transport property. Its subject has been continued to be attractive for researchers. Thus, we will examine in the present experiment a $Pd_{76}Cu_6Si_{18}$ glass with the chemical composition analogous to a well-known stable $Pd_{77.5}Cu_6Si_{16.5}$ glass, and a $Pd_{40}Ni_{10}Cu_{30}P_{20}$ glass with $\Delta T_x = T_x$-T_g $\approx$ 90 K, where the T_x and T_g are the crystallization and the glass transition temperatures, respectively and with a small critical cooling rate of amorphous formation $\approx$ 0.1 K/s. The stability of these glasses in the supercooled liquid region was examined in various atmospheres and it was concluded that the resistance change for these glasses was caused by the variation of electron transport associated with glass transition. A trial will be presented to interpret the behavior of electrical resistivity around glass transition temperature, based on Faber-Zimann theory. Another aspect in our present study is to investigate the behavior of hydrogen atom in the amorphous structure. As known well, Pd, Pt and Zr based crystalline alloys act as a hydrogen atom absorber. The Zr-Ni(Cu) amorphous alloy has been reported to show a similar feature [18]. However, the hydrogen atom absorption in Zr-Al-Ni and Pd-P glasses is not presented. We will report in present experiment on the anomalous behavior of electrical resistance, fairly likely to hydrogen atom absorption, in $Zr_{60}Al_{15}Ni_{25}$, $Pd_{76}Cu_6Si_{18}$ and $Pd_{40}Ni_{10}Cu_{30}P_{20}$ glasses. Especially, concerning to a $Pd_{40}Ni_{10}Cu_{30}P_{20}$ glass, the possibility of hydrogen atom absorption in the supercooled liquid region will be examined because this glass exhibited a significant small viscosity in this state [11].

3.2 Experimental Procedure

The procedure of sample preparation has been presented in previous papers [5,7] for a $Pd_{40}Ni_{10}Cu_{30}P_{20}$ and $Zr_{60}Al_{15}Ni_{25}$ glasses. Master alloys of nominal composition $Pd_{82}Si_{18}$ and $Pd_{76}Cu_6Si_{18}$ were produced by arc-melting. Then, to reduce the possible composition heterogeneity of the alloy due to precipitation, the melting operation was repeated five times as turning the ingot. The amorphous alloy in a ribbon form was prepared by a single-roller melt spinning technique. The quenching operation was performed almost in an Ar atmosphere. Only $Zr_{60}Al_{15}Ni_{25}$ glass was produced also under vacuum (1×10^{-4} Torr). Amorphous ribbons possessed the cross sections of 0.02×10, 0.03×2 (in an Ar atmosphere) or 0.06×1 (in vacuum), 0.04×2 and 0.03×2 mm^2 for a $Pd_{40}Ni_{10}Cu_{30}P_{20}$, $Zr_{60}Al_{15}Ni_{25}$, $Pd_{82}Si_{18}$ and $Pd_{76}Cu_6Si_{18}$ glasses. The amorphicity of an as-quenched sample was examined by X-ray diffraction and transmission electron microscopy (TEM). The amorphous phase transition such as glass transition and crystallization was measured by differential scanning calorimetry (DSC) mostly with a heating rate of 0.67 K/s. Then the onset of glass transition and crystallization was defined as a departure from the base line on DSC curve. The electrical resistance measurement was carried out by usual four-probe technique only with a heating rate of 0.67 K/s, where the infrared image furnace was employed to realize a relative high heating rate. The different gas atmospheres, that is, Ar (99.9995%), He (99.9999%), H_2 (99.99999%) and vacuum (5.0×10^{-6} Torr) were used to investigate the influence of oxidation and hydrogen atom absorption on the electrical resistance behavior.

3.3 Results and Discussion

3.3.1 Pd-Si Based Glasses

The X-ray diffraction pattern for an as-quenched $Pd_{82}Si_{18}$ glass is shown in Fig. 3.1 together with one for the sample crystallized by heating to 823 K in an Ar atmosphere. This profile shows that an as-quenched sample can be regarded as a single amorphous phase. The thermal behavior of a $Pd_{82}Si_{18}$ glass was investigated by the electrical resistance measurement under various atmospheres and DSC scan with a heating rate of 0.33 K/s in an Ar atmosphere. The results are shown in Fig. 3.2, where the resistance (left side axis) is scaled by the initial value at room temperature and the heat flow (arbitrary unit) in DSC measurement is shown in the right side axis. The features for these electrical resistance and DSC curves are summarized as follows. The electrical resistance curve possesses a positive temperature coefficient of resistance around room temperature and increases with increasing the temperature. The supercooled liquid region is distinctly observed in DSC curve. The glass transition temperature T_g (630 K) is in fairly good

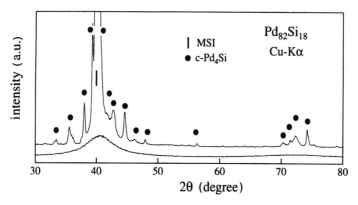

Fig. 3.1. X-ray diffraction patterns of a $Pd_{82}Si_{18}$ glass for as-quenched (lower) and crystallized (upper) states (heated to 823 K)

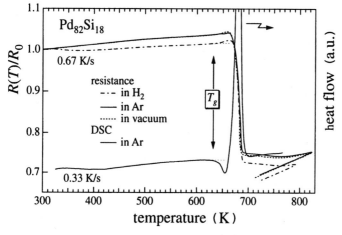

Fig. 3.2. The behavior of electrical resistance in a $Pd_{82}Si_{18}$ glass in Ar, H_2 or vacuum, together with DSC curve measured in an Ar atmosphere

agreement with electrical resistance and DSC measurements, but the crystallization temperature T_x is somewhat lower for DSC measurement than that for the electrical resistance measurement, corresponding to the employment of the slower heating rate for DSC measurement. The slope of electrical resistance curves increases clearly together with the onset of glass transition for all curves in spite of the difference in the atmosphere used. The similar behavior has been found in $Pd_{80}Si_{20}$ glass and attributed to the formation of tiny particles in the supercooled liquid region [6]. X-ray diffraction patterns were taken at room temperature for samples heated to 650 K in Ar or H_2 atmosphere to examine this fact and the result is shown in Fig. 3.3. The sharp diffraction peaks are observed in a lower side of principal amorphous

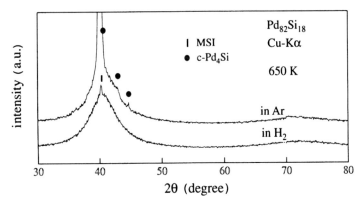

Fig. 3.3. X-ray diffraction patterns for $Pd_{82}Si_{18}$ samples heated to 650 K in the supercooled liquid region under an Ar or H_2 atmosphere

halo pattern. This suggests strongly the precipitation of nano-scaled particles similarly in this glass. Also, the peak intensity is significantly stronger for an Ar atmosphere than that for a H_2 one. This implies that the oxygen atom plays an important role on the precipitation of crystallite. The electrical resistance curve in a H_2 atmosphere showed an initial raise at room temperature, followed by a minimum at about 350 K with progressive heating. This may be attributed presumably to hydrogen atom absorption and desorption, respectively.

It is found in the present experiment that the electrical resistance increases with time at room temperature. As-quenched sample was heated to about 420 K in vacuum to reduce the thermal history. After cooled to room temperature, the hydrogen gas was introduced into the apparatus and the electrical resistance measurement was started under a constant pressure of about 0.12 MPa. In addition, the similar experiment was performed for the sample crystallized by heating to 823 K in vacuum. The result was shown in Fig. 3.4. The resistance value of amorphous sample increases initially with time and reaches a saturation level (about 2.8 %) after a duration of about 5.5×10^2 s. However, the crystallized sample does not exhibit such behavior. The slight decrease in resistance for crystallized sample corresponds to a gradual decrease of temperature after introduction of hydrogen gas into the apparatus because the electrical resistance of the crystallized phase showed a large positive value of temperature coefficient. The crystallized phase was examined by X-ray diffraction and most peaks in the pattern were attributed to a metastable c-Pd_4Si (code number 24-819 of ASTM card). A very strong unknown peak was observed at a 2θ position corresponding to $d=0.226$ nm (a lattice constant $a=0.391$ nm assuming the cubic structure). By using diffraction data of a $Pd_{80}Si_{20}$ glass reported previously [12], this unknown peak was attributed to a metastable MSI phase (a solid solution of Pd-Si with $a=0.400$ nm).

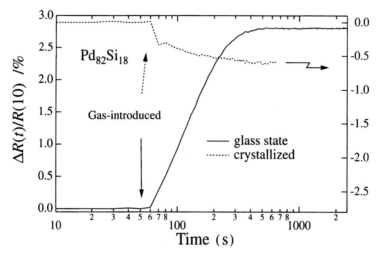

Fig. 3.4. The time change in the electrical resistance for as-quenched and crystallized $Pd_{82}Si_{18}$ samples after introduction of hydrogen gas into the apparatus at room temperature

It is known well that replacement of some Pd in Pd-Si glass with a few % Cu enhances the stability of this glassy system [14]. We examined the electrical resistance behavior of glassy $Pd_{76}Cu_6Si_{18}$ alloy in various atmospheres. Figure 3.5 shows the result together with DSC thermogram. The T_g (600 K) and T_x (673 K) in both experiments are in good agreement with each other. The behavior of electrical resistance curve is analogous to that

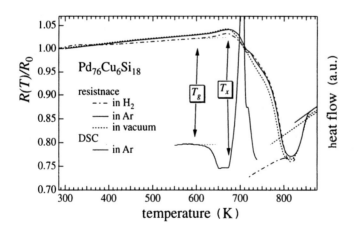

Fig. 3.5. The behavior of electrical resistance in a $Pd_{76}Cu_6Si_{18}$ glass in Ar, H_2 or vacuum

of $Pd_{82}Si_{18}$ glass. To check whether or not the resistance increment in the supercooled liquid region is attributed similarly in this case to the precipitation of any small crystallites. TEM micrograph was taken for the sample heated to 670 K in an Ar atmosphere and the result is shown in Fig. 3.6. No

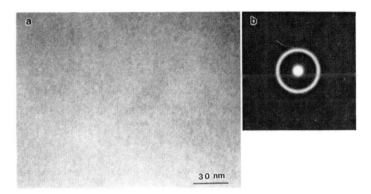

Fig. 3.6. (a) Transmission electron micrograph and (b) diffraction pattern for a $Pd_{76}Cu_6Si_{18}$ glass heated to 670 K just below T_x in an Ar atmosphere

crystallites are visible in the micrograph and corresponding diffraction pattern shows only diffuse halo rings peculiar to an amorphous phase. Therefore, we conclude that this metallic glass can be regarded as a single amorphous phase in the supercooled liquid region at least under dynamical heating at a conventional rate. The change in the slope of electrical resistance curve after glass transition suggests strongly that the electron transport property is varied after glass transition. We have reported the similar behavior of electrical resistance in a $Zr_{60}Al_{15}Ni_{25}$ metallic glass after glass transition [9]. But, in this alloy system, it was difficult to interpret the change in the electron transport property only from glass transition because of an easily oxidizable tendency of this alloy system. The slope of electrical resistance curves in the supercooled liquid region for the samples measured in Ar or H_2 atmosphere shows almost the same value (2.0×10^{-4} K^{-1}) as that in vacuum. This means that a $Pd_{76}Cu_6Si_{18}$ metallic glass is almost unaffected by the oxide impurity in the atmosphere. Like a $Pd_{82}Si_{18}$ glass, the electrical resistance increased with time at room temperature in a H_2 atmosphere. Figure 3.7 shows the time dependence of electrical resistance in a H_2 atmosphere for as-quenched and crystallized samples. The electrical resistance curve of amorphous sample reached a saturation level (about 2.6 %) like a $Pd_{82}Si_{18}$ glass after a time duration of about 10^3 s. In contrast to a $Pd_{82}Si_{18}$ glass, however, the crystallized sample exhibited the increment of electrical resistance in a H_2 atmosphere as shown in Fig. 3.7. X-ray diffraction pattern is shown in Fig. 3.8 for the samples which was heated to 655 K and crystallized by heating to 830 K in an Ar atmosphere. In present glass, a c-Pd_9Si_2 (orthorhombic,

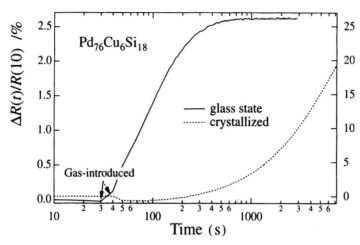

Fig. 3.7. The time change in the electrical resistance for an as-quenched and crystallized $Pd_{76}Cu_6Si_{18}$ samples after introduction of hydrogen gas into the apparatus at room temperature

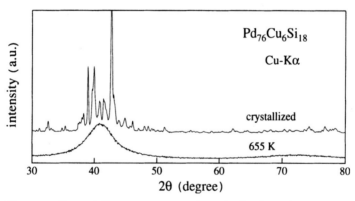

Fig. 3.8. X-ray diffraction patterns for $Pd_{76}Cu_6Si_{18}$ glasses in amorphous (heated to 655 K) and crystallized (heated to 830 K) states

code number 41-1102 of ASTM card), a c-Pd_5Si (Monoclinic, code number 35-1241), the same MSI phase as $Pd_{82}Si_{18}$ glass and unknown phase were visible after crystallization. No copper silicide phases were observed in our present experiment. This may imply that a small fraction of Cu was dissolved into the forms of a c-(Pd, Cu)$_9$-Si$_2$ and c-(Pd, Cu)$_5$Si. It has been reported [10] that the alloy addition of Ni has significantly influenced on the crystallization process of a Pd-Si glass. Therefore, it is acceptable even if crystallized phases for a $Pd_{76}Cu_6Si_{18}$ glass are different from those for a $Pd_{82}Si_{18}$ glass. Although the similar Pd-rich fcc phase was detectable in both of a $Pd_{82}Si_{18}$ and $Pd_{76}Cu_6Si_{18}$ glasses, the raise of electrical resistance, presumably due to hydrogen atom absorption, was not observed for a $Pd_{82}Si_{18}$

glass. It has been reported that the hydrogen atom absorption is significantly dependent on the surface state of the alloy, especially a presence of the oxide layer [15]. Therefore, the surface state of these glasses after crystallization should be examined in details by the method such as SEM observation and Auger electron spectroscopy.

3.3.2 $Pd_{40}Ni_{10}Cu_{30}P_{20}$ Glass

Replacement of Ni atom in a $Pd_{40}Ni_{40}P_{20}$ glass with 30 at % Cu has improved significantly the stability and the glass forming ability for this metallic glass [5]. In addition, the electrical resistance curve has been reported to show a negative value of temperature coefficient over all temperature range containing the supercooled liquid region [14]. In the present experiment, the behavior of electrical resistance in a H_2 atmosphere was compared to that in a He atmosphere. Four strips with a form of 1.5 mm × 10 mm were cut from a ribbon sample. Electrical resistance experiments were twice repeated under the same experimental condition for respective atmospheres employed. The result is presented in Fig. 3.9. The T_g (570 K) and T_x (660 K) for all electrical

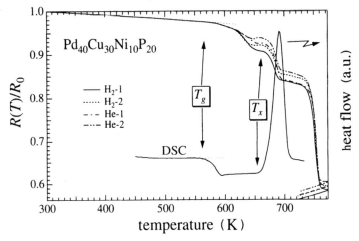

Fig. 3.9. The temperature dependence of electrical resistance curves measured in He or H_2 atmosphere for a $Pd_{40}Ni_{10}Cu_{30}P_{20}$ glass. The reproducibility of curves for two samples was fairly good in each atmosphere

resistance measurements are in agreement with that for DSC measurement. The change in the slope of electrical resistance curve is observed clearly over T_g. The reproducibility of electrical resistance curves is in excellent agreement with each other until about 620 K. From this temperature to 640 K, the curves in a H_2 atmosphere deviate from those in a He atmosphere. However, the reproducibility of two curves in each atmospheres is really acceptable. No crystallites have been observed in TEM micrograph for the sample heated to 640 K in an Ar atmosphere [11]. The sample was heated to 630 K in a H_2 atmosphere, followed by subsequent cooling to room temperature, and the TEM micrograph was taken at room temperature. The result is shown in Fig. 3.10. The micrograph and corresponding diffraction pattern reflects

Fig. 3.10. Transmission electron micrograph and diffraction pattern for a $Pd_{40}Ni_{10}Cu_{30}P_{20}$ glass heated to 630 K in the supercooled liquid region under a H_2 atmosphere

only the amorphous nature of this sample. Therefore, we conclude that the reduction of electrical resistance immediately over T_g corresponds intrinsically to the change in electron transport due to glass transition. In addition, a subsequent electrical resistance drop between 620 and 640 K seen under H_2 atmosphere may be attributed, very likely, to hydrogen atom absorption. The reproducibility of electrical resistance curve is not good between 640 and T_x. We were noticed that the surface of the strip became wavy and bumpy significantly beyond 640 K. The similar change in the ribbon form with progressive annealing has been reported in a $Pd_{40}Ni_{40}P_{20}$ glass [7]. It influenced presumably on the shape factor in the electrical resistance formula, leading to a scatter of resistance value in this region. It has been reported [11] that the viscosity of this glass was significantly reduced in the supercooled liquid region, and the minimum viscosity was only 8.3×10^5 Pa/s. This is about two order of magnitude lower than 10^8 Pa/s reported for a $Pd_{77.5}Cu_6Si_{16.5}$ glass [13]. As known well, the equilibrium free volume in the amorphous material is proportional to the inverse of a logarithm of the viscosity [16]. Therefore, it is presumed that the local structure of a $Pd_{40}Ni_{10}Cu_{30}P_{20}$ glass in the supercooled liquid state was transformed into which the hydrogen atom penetrated easily. On the other hand, in a $Pd_{76}Cu_6Si_{18}$ glass, the slope of electrical resistance was almost unchanged in the supercooled liquid region in spite of different gas atmospheres employed. This may mean that a $Pd_{76}Cu_6Si_{18}$ glass is hard to absorb hydrogen atom in the supercooled liquid region. Of cause, a distinct evidence whether or not the hydrogen atom absorption takes place in the supercooled liquid is not proposed in our present experiment. The further examination is necessary to corroborate our interpretation for some anomalous behaviors of electrical resistance seen in these glasses.

3.3.3 $Zr_{60}Al_{15}Ni_{25}$ Glass

Annealing behavior of electrical resistance for a typical Zr-based glass, $Zr_{60}Al_{15}Ni_{25}$, was examined in Ar, H_2 or vacuum. The result is presented in Fig. 3.11. Within the glassy state region below glass transition temperature, all curves show an almost similar temperature dependence. However, the resistance in Ar or H_2 atmosphere showed a larger value, compared with that in vacuum, in the supercooled liquid region. The high resistance value in an Ar atmosphere has been attributed to the influence of oxide impurity in an Ar gas by means of Auger spectroscopy [7]. We employed a highly pure H_2 gas in present experiment. So this phenomenon under H_2 gas atmosphere is very likely attributed to hydrogen atom absorption in the supercooled liquid region. When the sample has been heated to 750 K in a middle of the supercooled liquid region under an Ar atmosphere, no crystallites have been observed in TEM micrograph taken at room temperature [7]. Corresponding micrograph was taken in a H_2 atmosphere and presented in Fig. 3.12. Similarly, no clear evidence for crystallization is found. To confirm the hydrogen atom absorption into this glass, we performed experiments using

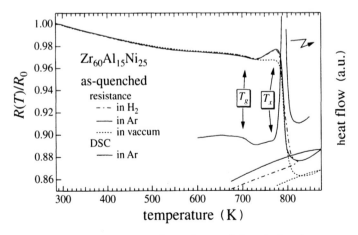

Fig. 3.11. The temperature dependence of electrical resistance curves measured in Ar, H_2 or vacuum for a $Zr_{60}Al_{15}Ni_{25}$ glass, which was prepared in an Ar atmosphere

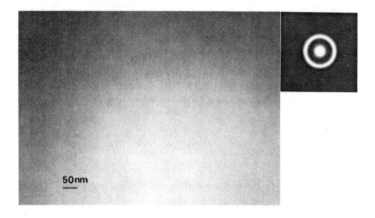

Fig. 3.12. Transmission electron micrograph and diffraction pattern for a $Zr_{60}Al_{15}Ni_{25}$ glass heated to 750 K in the supercooled liquid region under a H_2 atmosphere

the sample prepared in vacuum and one from which the surface oxide was removed by mechanical polishing. Figure 3.13a shows comparatively the electrical resistance curves measured in different atmospheres for polished samples, and Fig. 3.13b presents the difference of electrical resistance behavior in a H_2 atmosphere for the samples prepared in Ar or vacuum. As seen clearly, the electrical resistance exhibits a fairly higher value in H_2 than in Ar over all temperature range. It has been pointed out that the electrical resistance was increased by hydrogen atom absorption for a Zr-based metallic glass [17–19]. It seems that the hydrogen atom penetrated through the cracks in the oxide layer which was produced by large thermal expansion of the supercooled liquid [20]. Also, the removal of oxide layer from the surface area is predicted to enhance the efficiency of hydrogen atom absorption [15]. Therefore, although the characterization is not sufficient, these facts are suggestive of hydrogen atom absorption enhanced in the supercooled liquid region.

3.3.4 Change in Electrical Resistivity Associated with Glass Transition

We examined in detail the behavior of electrical resistance associated with glass transition for stable $Pd_{76}Cu_6Si_{18}$ and $Pd_{40}Ni_{10}Cu_{30}P_{20}$ glasses. Fig. 3.14 shows the electrical resistance curves versus annealing temperature nearby T_g in $Pd_{76}Cu_6Si_{18}$ glass. Then four running curves were measured by the repetition of heat treatment between 582 K in a glassy state and 656 K in a supercooled liquid state across the glass transition temperature with a heating and cooling rate of 0.33 K/s in an Ar atmosphere. In each curve, it is shown that the slope $dR(T)/dT/R(300)$ increases always

3 The Anomalous Behavior of Electrical Resistance   81

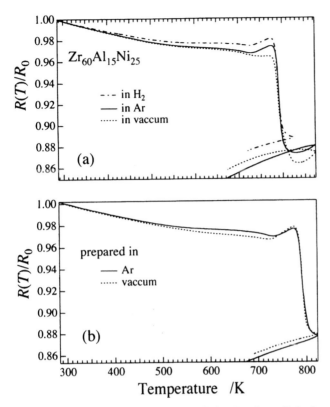

Fig. 3.13. (a) Electrical resistance behavior for polished samples, where the measurements were performed in Ar, H_2 or vacuum. (b) Electrical resistance curves measured in a H_2 atmosphere for samples prepared in an Ar atmosphere or vacuum

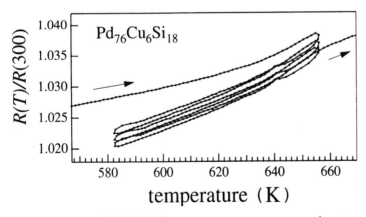

Fig. 3.14. The $R(T)/R(300)$ versus temperature across glass transition temperature

after glass transition. This suggests that the crystallization was possibly suppressed through the heat treatment because it causes usually the electrical resistance to decrease. Although the curve shifts slightly to a lower position as a whole after repetition of heat treatment, this may be attributed to an advance in the structural relaxation process. Similar behavior was observed for a $Pd_{40}Ni_{10}Cu_{30}P_{20}$ glass in repeating heat run between 540 and 600 K, and the result is presented in Fig. 3.15. In contrast to $Pd_{76}Cu_6Si_{18}$ glass, the

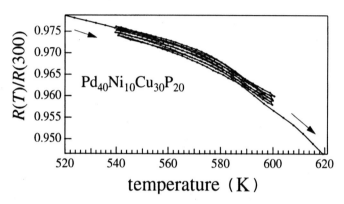

Fig. 3.15. $R(T)/R(300)$ versus temperature across glass transition temperature

slope $dR(T)/dT/R(300)$ shows the negative sign in all temperature regions. In addition, we can see that $dR(T)/dT/R(300)$ decreases further after glass transition. The repetition of heat treatment induces a lowering of curve similarly as $Pd_{76}Cu_6Si_{18}$ glass. The origin is presumed to be due to the structural relaxation. After this alternative heating and cooling operations, both samples heated continuously until 773 K to check the crystallization behavior. The almost same behavior as that for an as-quenched sample was observed for both glasses and this suggests that the crystallization in repeating heat treatment is presumed to be suppressed. It has been postulated [6] that the amorphous phase separation induced also the variation of electrical resistance in the supercooled liquid region. However, TEM images of $Pd_{76}Cu_6Si_{18}$ and $Pd_{40}Ni_{10}Cu_{30}P_{20}$ samples heated to a temperature in the supercooled liquid region did not show a coarse texture peculiar to the amorphous phase separation [6]. Two step endothermic reaction, suggesting two step glass transition, was undetected in DSC thermogram. Thus, it is concluded that the variation in the slope $dR(T)/dT/R(300)$ after glass transition reflected the electron transport property peculiar to supercooled liquid state. The linear thermal expansion coefficient of $Pd_{40}Ni_{10}Cu_{30}P_{20}$ glass has been measured recently in the glassy solid region and liquid region [21] by means of specific volume expansion experiment.

The value has been reported to be the order of 10^{-5} K^{-1} in both regions. Although the value in the supercooled liquid region was not obtained for the difficulty in the experiment, it may be deduced to have the same order from interpolation of the relation between the specific volume and the temperature in that experiment. Therefore, assuming the thermal expansion coefficient of $Pd_{76}Cu_6Si_{18}$ glass is the same order as that for $Pd_{40}Ni_{10}Cu_{30}P_{20}$ glass, the contribution to the electrical resistance from thermal expansion of the sample during annealing can be neglected for these glasses. So, instead of the electrical resistance, we can use the electrical resistivity in discussing the variation in electron transport associated with glass transition. The vertical scale $R(T)/R(300)$ in Fig. 3.14 and 3.15 can be approximately saved as $\rho(T)/\rho(300)$. The room temperature resistivity of $Pd_{76}Cu_6Si_{18}$ and $Pd_{40}Ni_{10}Cu_{30}P_{20}$ glasses were estimated to be 0.81±0.04 and 2.33±0.2 μΩm, respectively. The temperature coefficient of resistivity (TCR) at room temperature $(d\rho/dT/\rho)_{300}$ showed the positive value 8.32×10^{-5} K^{-1} for $Pd_{76}Cu_6Si_{18}$ glass and the negative one -7.90×10^{-5} K^{-1} for $Pd_{40}Ni_{10}Cu_{30}P_{20}$ glass. This indicates that these glasses satisfy the Mooijs' relation [22], which predicts that $(d\rho/dT/\rho)_{300}$ changes from a positive to a negative value when the room temperature resistivity increases roughly over 1.50 μΩm. The change in the electrical resistivity after glass transition can be interpreted qualitatively by assuming the validity of Faber-Zimann theory for the electron transport property of these alloy system. Following this formalism, the sign of $d\rho/dT$ can be explained by considering the change in the shape of structure factor $S(Q)$ with increasing temperature, and the relative position of $2k_F$ (k_F means the Fermi wave number of the conduction electron) to Q_p (the first peak position in $S(Q)$ profile) [23]. Because ρ is given by integral over $(Q/2k_F)$-space from 0 to unity and the integrand contains the scattering potential for conduction electron, $S(Q)$ and the term $(Q/2k_F)^3$. The temperature dependence of ρ is taken into account through $S(Q)$ in this case. When the temperature increases, the peak intensity in $S(Q)$ can be expected to decrease and $S(Q)$ at lower Q values tends to rise, on the contrary, due to the increase of vibrational amplitude of atom. So if $2k_F$ is located at a position in lower Q side, ρ can be expected to increase with temperature. Otherwise, if $2k_F$ is nearby Q_p, ρ can be expected to fall with temperature. The Q_p for a $Pd_{76}Cu_6Si_{18}$ and $Pd_{40}Ni_{10}Cu_{30}P_{20}$ glasses were examined by X-ray diffraction and estimated to be 28.4 and 29.0 nm^{-1}, respectively. The $2k_F$ for a $Pd_{77.2}Cu_6Si_{16.8}$ glass has been estimated to be 24.8 nm^{-1} by means of Hall coefficient experiment [24]. $S(Q)$ for $Pd_{76}Cu_6Si_{18}$ glass was below unity until $Q=25.0$ nm^{-1}. Thus, $2k_F$ may be predicted to locate at a lower Q side for this glass analogous to the composition of $Pd_{77.2}Cu_6Si_{16.8}$ glass. The Hall coefficient data for $Pd_{40}Ni_{10}Cu_{30}P_{20}$ glass has not been reported yet. So we calculated $2k_F$ using the expression $k_F = (3\pi^2 n_0)^{1/3}$, where the mean conduction electron density n_0 per unit volume is defines as $n_0 = d_0 \sum c_i n_i$

with mean atomic number density d_0, concentration of element c_i and effective conduction electrons per atom n_i. Here, it is assumed that P atoms are in fully ionized five-valent state and $n_{Cu}=1.0$. For Ni atoms, $n_{Ni}=0.88$ per atom was used [25]. We also used $n_{Pd}=0.36$, the value in Pd crystal [26]. In these cases, it was postulated that part of valence electrons were transferred to d-band state and effective conduction electrons were decreased. The d_0 was estimated to be 77.2 nm^{-3} from the density 9.33 gcm^{-3} measured by Archimedes' method with toluene as a working fluid. Evidently, $2k_F=30.4$ nm^{-1} was obtained for this glass. Thus, the relation $2k_F \approx Q_p$ is approximately satisfied for $Pd_{40}Ni_{10}Cu_{30}P_{20}$ glass. When the glass transition occurs, it will be expected that the glass structure becomes more liquid-like, leading to further decreasing of the peak intensity and broadening of peak width in $S(Q)$. Meanwhile, at lower values of Q, $S(Q)$ can be expected further to increase with increasing the temperature. Therefore, ρ is expected further to decrease after glass transition for $Pd_{40}Ni_{10}Cu_{30}P_{20}$ glass and increase for $Pd_{76}Cu_6Si_{18}$ glass. These give only qualitative explanation about the increase and the decrease of the resistivity for $Pd_{76}Cu_6Si_{18}$ and $Pd_{40}Ni_{10}Cu_{30}P_{20}$ glasses after glass transition. The convincing explanation of this observation will be offered by means of in-situ diffraction experiment carried out in the supercooled liquid region.

3.4 Concluding Remarks

The electrical resistance behavior was examined for metallic glasses $Pd_{82}Si_{18}$, $Pd_{76}Cu_6Si_{18}$, $Pd_{40}Ni_{10}Cu_{30}P_{20}$ and $Zr_{60}Al_{15}Ni_{25}$ in an Ar, H_2 atmosphere or vacuum. The electrical resistance increased with time at room temperature in $Pd_{82}Si_{18}$ and $Pd_{76}Cu_6Si_{18}$ glasses in H_2 atmosphere, and subsequent dynamical annealing resulted in a minimum around 350 K on the electrical resistance curve, possibly due to hydrogen atom absorption and desorption. The c-Pd-like crystallites in a $Pd_{82}Si_{18}$ glass, formed in an early crystallization stage, did not show the change in electrical resistance at room temperature under a H_2 atmosphere. A $Pd_{40}Ni_{10}Cu_{30}P_{20}$ glass exhibited the slight decrease of electrical resistance in the supercooled liquid region in H_2 atmosphere. The behavior of electrical resistance under a H_2 atmosphere in a $Zr_{60}Al_{15}Ni_{25}$ glass was strongly influenced by the surface state of the sample. That is, the sample prepared in an ordinary Ar atmosphere showed a higher resistance value only in the supercooled liquid region. On the other hand, the sample prepared in vacuum and polished one exhibited the raise of electrical resistance over almost all temperature region. The change in the slope of electrical resistance curve was clearly observed over T_g for stable $Pd_{76}Cu_6Si_{18}$ and $Pd_{40}Ni_{10}Cu_{30}P_{20}$ glasses. Considering that no clear evidence for nanocrystallization and amorphous phase separation was discovered at least within X-ray diffraction and TEM observation, this phenomenon is very likely attributed to an essential variation of electron transport property associated with glass transition. The mechanism of this variation may be interpreted qualitatively by assuming that the electron transport of these alloy systems is described by Faber Zimann theory.

References

1. A. Inoue, T. Zhang and T. Masumoto: Zr-Al-Ni amorphous alloys with high glass transition temperature and significant supercooled liquid region. Mater. Trans. JIM **31**, 177-183 (1990)
2. A. Peker and W.L. Johnson: A highly processable metallic glass: $Zr_{41.2}Ti_{13.8}Cu_{12.5}Ni_{10.0}Be_{22.5}$. Appl. Phys. Lett. **63**, 2342-2344 (1990)
3. A. Inoue, M. Kohinata, A.P. Tsai and T. Masumoto: Mg-Ni-La amorphous alloys with a wide supercooled liquid region. Mater. Trans. JIM **30**, 378-381 (1989)
4. A. Inoue, T. Zhang and T. Masumoto: Al-La-Ni amorphous alloys with a wide super cooled liquid region. Mater. Trans. JIM **30**, 965-972 (1989)
5. A. Inoue, N. Nishiyama and T. Matsuda: Preparation of bulk $Pd_{40}Ni_{10}Cu_{30}P_{20}$ alloy of 40 mm in diameter by water quenching. Mater. Trans. JIM **37**, 181-184 (1996)
6. H.S. Chen and D. Turnbull: Formation, Stability and structure of palladium-silicon based alloy glasses. Acta Metal. **17**, 1021-1031 (1971)
7. O. Haruyama, H.M. Kimura and A. Inoue: Thermal stability of Zr-based glassy alloys examined by electrical resistance measurement. Mater. Trans. JIM **37**, 1741-1747 (1996)
8. O. Haruyama, H. Kimura, T. Aoki, N. Nishiyama and A. Inoue: Thermal stability of Pd-based metallic glasses examined by electrical resistance measurement. Sci. Rep. RITU A **43**, 97-100 (1997)
9. R. Brüning: Structural relaxation and the glass transition in metallic glasses. A doctoral thesis, McGill University, 1990
10. Y. Lanping and H. Yizhen: Correlation between the microstructure and internal friction peaks of metallic glass $Pd_{77.5}Ni_{6.0}Si_{16.5}$. J. Non-Cryst. Solids **105**, 33-38 (1988)
11. N. Nishiyama and A. Inoue: Glass transition behavior and viscous flow working of $Pd_{40}Ni_{10}Cu_{30}P_{20}$ amorphous alloy. Mater. Trans. JIM, received
12. T. Masumoto and R. Maddin: The mechanical properties of palladium 20 a/o silicon alloy quenched from the liquid state. Acta Metal. **19**, 725-741 (1971)
13. G. Yiqin, Z. Fuqian, L.Xiong and L. Zongyao: Viscosity of metallic glasses $PdCu_6Si_{16.5}$, $PdNi_6Si_{16.5}$ and $CuZr_{43}$ and their activation energies of viscous flow. Acta Met. Sin. **18**, 696-701 (1982)
14. H. Kimura, A. Inoue, N. Nishiyama, K. Sasamori, O. Haruyama and T. Masumoto: Thermal, mechanical and physical properties of supercooled liquid in Pd-Cu-Ni-P. Sci. Rep. RITU A **43**, 101-106 (1997)
15. H. Uchida, Y. Ohtani, M. Ozawa, T. Kawahata and T. Suzuki: Surface processes of H_2 in the initial activation of $LaNi_5$. J. Less-Common Met. **172/174**, 983-996 (1991)
16. A.K. Doolittle: Studies in Newtonian flow. II. The dependence of the viscosity of liquids on free-space. J. Appl. Phys. **22**, 1471-1475 (1951)

17. Y. Yamada, Y. Itoh, T. Matsuda and U. Mizutani: Electron transport studies of $Ni_{33}Zr_{67}$-based metallic glasses containing H, B, Al and Si. J. Phys. F:Met. Phys. **17**, 2313-2322 (1987)
18. J. Garaguly, A. Lovas, . Cziriki, M. Reybold, J. Takics, K. Wetzig: Reversible and irreversible hydrogen absorption in $Ni_{67-x}Cu_xZr_{33}$ glasses monitored by in situ resistivity measurements. Mater. Sci. and Eng. A **226-228**, 938-942 (1997)
19. G. PetŽ, I. Bakonyi, K. Tompa and L. Guczi: Photoemission investigation of the elctronic-structure changes in Zr-Ni-Cu metallic glasses upon hydrogenation. Phys. Rev. B **52**, 7151-7158 (1995)
20. H. Kimura, M. Kishida, T. Kaneko, A. Inoue and T. Masumoto: Physical and Mechanical Properties of Zr-Based Metallic Glasses. Mater. Trans. JIM **36**, 890-895 (1995)
21. M. Horino, N. Nishiyama, Y. Yokoyama and A. Inoue: unpublished research
22. J. H. Mooij: Electrical Conduction in Concentrated Disordered Transition Metal Alloys. Phys. Stat. Sol.(a) **17**, 521-530 (1973)
23. J. S. Dugdale: The electrical properties of disordered metals, (Cambridge University press 1995)
24. U. Mizutani and T. B. Massalski: Hall-effect measurements and the electronic structure of amorphous Pd-Si-(Cu) alloys. Rhys. Rev. B **21**, 3180-3183(1980)
25. S. S. Jaswal: Electronic structure and properties of transition-metal-metalloid glasses: $Ni_{1-x}P_x$. Phys. Rev. B **34**, 8937-8940 (1986)
26. A. P. Cracknell: The Fermi Surface. II. d-block and f-block Metals. Adv. Phys. **20**, 1-141 (1971)

4 Methods for Production of Amorphous and Nanocrystalline Materials and Their Unique Properties

T. Aihara[1], E. Akiyama[1], K. Aoki[2], M. Sherif El-Eskandarany[3],
H. Habazaki[1], K. Hashimoto[1], A. Kawashima[1], M. Naka[4], Y. Ogino[5],
K. Sumiyama[1], K. Suzuki[1], and T. Yamasaki[5]

[1] Institute for Materials Research, Tohoku University, Sendai, 980-8577 Japan
[2] Department of Materials Science, Kitami Institute of Technology, Kitami 090, Japan
[3] Mining and Petroleum Engineering Department, Faculty of Engineering, Al-Azhar University, 11884 Nasr City, Cairo, Egypt
[4] Joining and Welding Research Institute, Osaka University Mihogaoka 11-1, Ibaraki, Osaka 567, Japan
[5] Department of Materials Science and Engineering, Faculty of Engineering, Himeji Institute of Technology, 2167 Shosha, Himeji, Hyogo 671-2201, Japan

Summary. There are a lot of possible techniques that can be used to prepare amorphous materials. This Chapter is denoted to the three widely used methods of preparation of amorphous and metastable materials namely: ball milling, electrodeposition and sputtering. The properties of materials obtained by these methods are also to be discussed in the present Chapter. Among the materials studied are: amorphous and nanocrystalline Co-Ti, Ni-W alloys containing from 5 to 30 at.% W, Ti-B, Ti-Si, Ti-C, Ti-Al and Ni-Mo alloys. The structure and phase transformations in the above-mentioned alloys have been studied by means of X-ray diffraction, small angle X-ray scattering, high resolution transmission electron microscopy and differential thermal analysis.

4.1 Introduction

Since the amorphous phase is less thermodynamically stable (i.e. it possesses a greater free energy) than the corresponding crystalline phase the preparation of the amorphous materials can be regarded as the addition of excess free energy. Such materials require a special ways of production. There are lots of possible techniques that can be used to prepare amorphous materials. This Chapter is denoted to the three widely used methods of preparation of amorphous and metastable materials namely: ball milling, electrodeposition and sputtering. The properties of materials obtained by these methods are also to be discussed in the present chapter. Well known rapid solidification method has been briefly described in the previous Chapter. It is also very important to note that although many materials can be produced in an amorphous form by different techniques, the resulted amorphous phase not always have the same properties. This difference can be noticed as macroscopic structural inhomogeneity, the degree of short range order, an existence of medium range order and so on.

4.2 Crystalline-Amorphous Cyclic Transformation of Ball Milled $Co_{75}Ti_{25}$ Alloy Powder

4.2.1 Use of Mechanical Alloying Technique for Amorphization

Since 1983, the mechanical alloying (MA) method [1] has been widely employed for preparing several metallic amorphous alloys [2–7], using ball milling and/or rod milling techniques [8]. It has been reported that further milling of amorphous $Ti_{75}Al_{25}$ [9], $Al_{80}Fe_{20}$ [10], $Fe_{78}Al_{13}Si_9$ [11] and $Ti_{50}Al_{25}Nb_{25}$ [12] powders leads to amorphous-crystalline transformation (crystallization). Recently, the present authors have found a unique phenomenon, so called, cyclic crystalline-amorphous transformation during mechanical milling of binary metallic powders [13,14] using a high-energy ball mill. A single amorphous phase of $Co_{75}Ti_{25}$ alloy powders is obtained after a short MA time (11 ks). The obtained amorphous phase does not withstand against the impact and shear forces generated by the milling media and crystallizes into a metastable bcc-Co_3Ti phase after MA time of 86 ks. The bcc phase turns to become the same amorphous phase of $Co_{75}Ti_{25}$ upon milling for 360 ks. Similar to mechanical grinding [15], this phase transformation is attributed to the accumulation of lattice imperfections such as point and lattice defects, which raise the free energy and transformation from the crystalline phase (bcc-$Co_{75}Ti_{25}$) to a less stable phase (amorphous). Further milling leads to the formation of crystalline and/or amorphous phases depending on the MA time. This article reviews the characteristic features of the cyclic crystalline-amorphous transformations in $Co_{75}Ti_{25}$ powders by MA.

4.2.2 Ball Milling Procedure and Analyzing Technique

Pure elemental powders (99.9%) of Co (70 μm) and Ti (50 μm) were mixed to give the desired nominal composition of $Co_{75}Ti_{25}$ (at.%) in a glove box under a purified argon atmosphere and sealed in a stainless steel vial (SUS 316, 250 ml in volume) together with 50 stainless steel balls (SUS 316, 10 mm in diameter). The ball-to-powder weight ratio was maintained as 17:1. MA was performed in a high energy planetary ball mill (Fritsch P5) at a rotation speed of $4.2\,s^{-1}$. In order to avoid a temperature rise during the milling process, the MA experiment was stopped after every 1.8 ks of MA, when the vial temperature became about 320 K. After the vial temperature went down to about 300 K, the MA experiment was restarted for the long MA times of 11–720 ks. When the vial was only charged with the milling media and operated for 1.8 ks, the temperature reached to about 310 K. The MA process was performed under the same experimental conditions three times to confirm the reproducibility and to avoid any misleading results. The structural changes with the MA time of the powders were detected by X-ray diffraction (XRD) with Cu-Kα radiation and transmission electron microscopy (200 kV TEM/EDS). The samples were thermally analyzed with a differential thermal analysis (DTA) at a heating rate of 0.33 K/s. The magnetization of the milled powders was investigated with a vibrating sample magnetometer (VSM) at room temperature. An induction coupled plasma (ICP) emission method was used to analyze the concentration of Co and Ti, and the degree of Fe contamination in the milled powders. The gas contamination contents were determined by a He carrier fusion-thermal conductivity method.

4.2.3 Structural Changes vs. Milling Time

The XRD diffraction patterns of mechanically alloyed $Co_{75}Ti_{25}$ powder after selected milling times are shown in Fig. 4.1. The figure displays unexpected results of the phase transformations. In contrast to the initial mixture of polycrystalline hcp-Co and hcp-Ti (Fig. 4.1a), a broad diffuse and smooth halo appears after short milling time of 11 ks (Fig. 4.1b), suggesting the formation of an amorphous $Co_{75}Ti_{25}$ alloy. The XRD pattern of the sample milled for

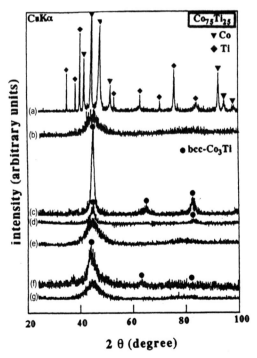

Fig. 4.1. XRD patterns of mechanically alloyed $Co_{75}Ti_{25}$ powders after (**a**) 0 ks, (**b**) 11 ks, (**c**) 86 ks, (**d**) 173 ks, (**e**) 360 ks, (**f**) 540 ks and (**g**) 720 ks of the MA time

11 ks then annealed well above its crystallization temperature (see Fig. 4.3) reveals an fcc structure. This amorphous phase is unstable against the impact and shear forces generated by the milling media (balls) and transforms into bcc-Co_3Ti upon milling for 86 ks, as shown in Fig. 4.1c. The reported equilibrium (fcc ordered) Co_3Ti phase has the $L1_2$ structure [16]. Therefore, this bcc-Co_3Ti phase is a metastable one and its lattice parameter is calculated to be 0.2855 nm, being smaller than that for the ordered bcc-CoTi equilibrium phase (0.2987 nm). After 173 ks of the MA time (Fig. 4.1d), the Bragg peaks for bcc-Co_3Ti become broader, suggesting the existence of an amorphous phase. After 360 ks of the MA time, this bcc-Co_3Ti phase transforms to the amorphous $Co_{75}Ti_{25}$ phase again, as illustrated in Fig. 4.1e. This phase transformation is attributable to the accumulation of imperfections of point and lattice defects, which raise the free energy from the metastable to a less stable state (amorphous). The MA time of 540 ks leads again to the formation of nanocrystalline bcc-Co_3Ti containing a small fraction of the amorphous phase, as presented in Fig. 4.1f. This bcc-Co_3Ti phase returns to the same amorphous $Co_{75}Ti_{25}$ phase after 720 ks of the MA time, as shown in Fig. 4.1g. Same transformations were observed for the other samples milled for three different milling runs. Thus, the sequence of the phase transformation with the MA time in ball-milled $Co_{75}Ti_{25}$ is as follows:

$$\begin{array}{cccccc} (11\,\text{ks}) & (86\,\text{ks}) & (360\,\text{ks}) & (540\,\text{ks}) & (720\,\text{ks}) \\ \text{Co+Ti} \rightarrow & \text{amorphous} \rightarrow & \text{bcc} \rightarrow & \text{amorphous} \rightarrow & \text{bcc} \rightarrow & \text{amorphous}. \end{array}$$

4.2.4 TEM Observations

TEM analyses using bright field images (BFI) and selected area diffraction patterns (SADP) have been used to understand the local structure of the milled powder during the different MA times. Figure 4.2 presents the BFI (a) and the corresponding SADP (b) of a sample milled for 11 ks of the MA

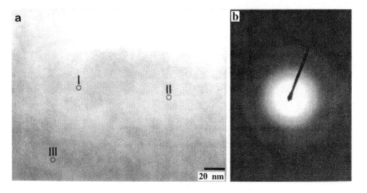

Fig. 4.2. BFI (a) and the corresponding SADP (b) of mechanically alloyed $Co_{75}Ti_{25}$ powders after 11 ks of the MA time

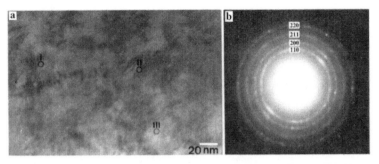

Fig. 4.3. BFI (**a**) and the corresponding SADP (**b**) of mechanically alloyed $Co_{75}Ti_{25}$ powders after 86 ks of the MA time

time. Overall, the sample appears to have a homogeneous fine structure with no dominant facet structure (Fig. 4.2a). The SADP shows a typical halo of an amorphous phase (Fig. 4.3b) in good agreement with the XRD pattern presented in Fig. 4.1b. In order to understand the local composition (within 3.5 nm) of the powder at this stage of milling, the sample is classified into three zones (I, II, III) for EDS investigations, as shown in Fig. 4.2a. The compositions of the sample at the selected zones are presented in Table 4.1. The sample is homogeneous and does not differ remarkably in compositions from one zone to another, indicating the formation of a homogeneous amorphous phase.

The BFI and corresponding indexed SADP of the powders milled for 86 ks of the MA time are presented in Fig. 4.3a and b, respectively. Numerous faults with grain boundary fringes and heavy dislocations appear in the powder as a result of the impact and shear stresses generated during MA, as shown in Fig. 4.3a. The SADP shows the formation of the bcc-Co_3Ti phase (see the Debye-Scherrer rings in Fig. 4.3b. No compositional difference could be detected between the samples milled for 86 and 11 ks: the average composition is still $Co_{75}Ti_{25}$, as presented in Table 4.1.

In the cyclic crystalline-amorphous transformation, the bcc-Co_3Ti phase returns an amorphous phase after 360 ks of the MA time. The formed amorphous phase has a fine and homogeneous structure (Fig. 4.4a) with a diffuse

Table 4.1. EDS analyses of ball milled $Co_{75}Ti_{25}$ powders after selected milling times

Milling time (ks)	Co content (at%)			Ti content (at%)		
	I	II	III	I	II	III
11	74.9	75.0	74.9	25.1	25.0	25.1
86	75.1	74.9	75.0	24.9	25.1	25.1
360	75.0	75.1	75.0	25.0	24.9	25.0

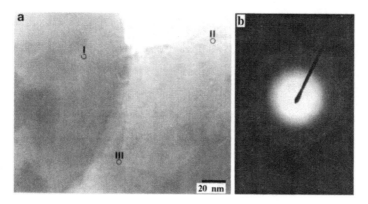

Fig. 4.4. BFI (a) and the corresponding SADP (b) of mechanically alloyed $Co_{75}Ti_{25}$ powders after 360 ks of the MA time

halo pattern, as presented in Fig. 4.4b. The average composition of the amorphous phase at these different zones (see Fig. 4.4a) is $Co_{75}Ti_{25}$, as displayed in Table 4.1.

4.2.5 Magnetization

Since the magnetic properties of Co-Ti alloys are sensitive to the structure change, we monitored the phase transformations with measuring the magnetization of mechanically alloyed $Co_{75}Ti_{25}$ powders. Figure 4.5 shows the magnetization of the milled powder at room temperature as a function of the MA time. The rapid decrease in the magnetization during the first few kiloseconds (3.6–7.2 ks) suggests a drastic decrease of pure Co particles in the mixture of $Co_{75}Ti_{25}$ powder and the formation of the amorphous phase. After 11 ks of the MA time, the magnetization decreases again to have a lower value of about 8.8×10^{-5} Wb·m/kg (70 emu/g). During the next stage of milling (14.4–22 ks), the magnetization of mechanically alloyed $Co_{75}Ti_{25}$ powders almost saturates at this value, suggesting the completion of the solid state amorphization reaction. After 43 ks of the MA time, the magnetization tends to increase to be of about 1.0×10^{-4} Wb·m/kg (80 emu/g), suggesting the formation of a ferromagnetic phase (bcc-Co_3Ti) in the amorphous matrix of $Co_{75}Ti_{25}$ alloy. After 86 ks of the MA time, the amorphous phase transforms completely into a crystalline phase of metastable bcc-Co_3Ti (see Fig. 4.1c) and the magnetization increases drastically to have a value of about 1.6×10^{-4} Wb·m/kg (125 emu/g). It is worth noting that this value is about two times larger than 7.0×10^{-5} Wb·m/kg (56 emu/g) for the ordered fcc-Co_3Ti phase obtained by annealing the amorphous $Co_{75}Ti_{25}$ phase at far above its crystallization temperature of 874 K. This indicates that the mechanically alloyed bcc-Co_3Ti phase has higher magnetization at room temperature. After 173 ks of MA time, the magnetization of bcc-Co_3Ti remarkably decreases due to the coexistence of amorphous $Co_{75}Ti_{25}$. Again, a crystalline-to-amorphous phase

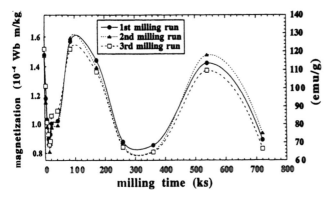

Fig. 4.5. Variation of the magnetization at room temperature for mechanically alloyed $Co_{75}Ti_{25}$ powders as a function of the MA time

transformation takes place during the next stage of milling (259–360 ks) so that the magnetization of the samples decreases to have almost the same values of the samples milled for 11–18 ks. After 540 ks of the MA time, the nanocrystalline bcc-Co_3Ti phase is formed and hence the magnetization increases again to have a value of about 1.4×10^{-4} Wb·m/kg (115 emu/g). This value is lower than that of the bcc-Co_3Ti powders milled for 86 ks, due to the coexistence of amorphous $Co_{75}Ti_{25}$ phase (see Fig. 4.1f). Up to the MA time of 720 ks, another crystalline-to-amorphous phase transformation is observed (Fig. 4.1g) and the yielded amorphous phase has almost the same magnetization as the one formed after the MA times of 18 ks and 360 ks (Fig. 4.5), confirming a cyclic phase transformations. Figures 4.1 and 4.5 indicate that the cyclic crystalline-amorphous transformation takes place periodically at about 360 ks of the MA time.

4.2.6 Thermal Stability

The DTA curves of mechanically alloyed $Co_{75}Ti_{25}$ powders are presented in Fig. 4.6 after selected MA times. The measurements were performed at a constant heating rate of 0.33 K/s under an argon gas atmosphere. All the samples were heated up to 1300 K (first run) and cooled down to about 400 K. Then, second heating runs (dashed lines) were performed in order to get the base line. After 11 ks of the MA time (Fig. 4.6a), a single sharp exothermic peak appears at 874 K. This exothermic reaction is attributed to the amorphous-crystalline phase transformation (crystallization). The XRD patterns of the sample heated up to 1300 K show the formation of the fcc-Co_3Ti phase. No remarkable change of the crystallization temperature (the exothermic peak temperature) can be detected in the sample milled for 22 ks. After 86 ks of the MA time, the exothermic reaction disappears and any

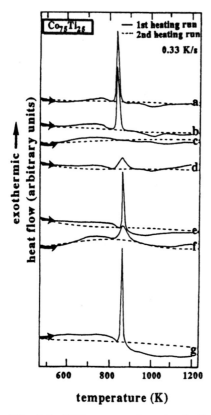

Fig. 4.6. DTA curves of mechanically alloyed $Co_{75}Ti_{25}$ powders milled after (a) 11 ks, (b) 22 ks, (c) 86 ks, (d) 173 ks, (e) 360 ks, (f) 540 ks and (g) 720 ks of the MA time

other reaction cannot be detected, as illustrated in Fig. 4.6c. This suggests the amorphous-crystalline (bcc-Co_3Ti) phase transformation of the as-milled sample. This crystalline phase is thermally stable and does not change to any other phase(s), as was confirmed by the XRD patterns of the sample heated up to 1300 K. However, it turns to the ordered fcc-Co_3Ti phase upon heating up to 1327 K. A broad exothermic peak at about 880 K is observed for the sample milled for 173 ks (Fig. 4.6d). This peak is due to the crystallization of a small fraction of the coexisting amorphous phase. The XRD patterns of this sample which heated up to 1300 K indicate the coexistence of fcc-Co_3Ti with the bcc-Co_3Ti phase formed already during MA. The bcc-Co_3Ti phase transforms to a single amorphous phase after 360 ks of the MA time, as suggested by the sharp exothermic reaction detected at 880 K in Fig. 4.6e. The crystallized sample reveals a structure of fcc-Co_3Ti phase. The broad exothermic peak of the sample milled for 540 ks indicates the existence of a small amount of the amorphous phase (Fig. 4.6f). The co-existence of bcc-Co_3Ti with fcc-Co_3Ti phases was confirmed by XRD of the sample heated up to 1300 K (far above this exothermic reaction). The bcc-Co$_3$Ti phase transforms to an amorphous bcc-Co$_{75}$Ti$_{25}$ phase upon milling for 720 ks and it crystallizes through a single exothermic reaction peak as shown in Fig. 4.6g.

4.2.7 Possible Reasons for the Cyclic Crystalline-Amorphous Transformations

One possible factor which may lead to such cyclic crystalline-amorphous transformations of Co$_{75}$Ti$_{25}$ is the existence of contamination in the ball-milled powder. Iron and gas (oxygen, nitrogen and hydrogen) contamination contents of the milled powders are plotted in Fig. 4.7 as a function of the milling time. During the first stage of MA, iron which comes from the stainless steel milling tool increases drastically to about 0.75 at.%. The gas contamination content that may be introduced to the milled powder during the ball-milling process and/or handling the sample outside the glove box increases during this stage of milling. This is attributed to the formation of active-fresh surfaces of the elemental powders (in particular Ti), which are able to react fast with the surrounding atmosphere. As the MA time increases, the milling tools are coated with the milled powder. This coating plays an important role to prevent introducing a further iron contamination to the milled powder. The iron contamination content of the end-product is about 0.80 at.%, as illustrated in Fig. 4.7. In addition, no remarkable changes in the gas content can be detected for the samples that milled for longer times. These contamination contents are almost negligible in the milled powder and can not be considered to be responsible for the cyclic phase transformations of Co$_{75}$Ti$_{25}$ during the ball milling process. The another possible factor which can lead to the phase transformations is the excess heating during a MA. To avoid such a factor, the ball-milling process was interrupted when the temperature of the

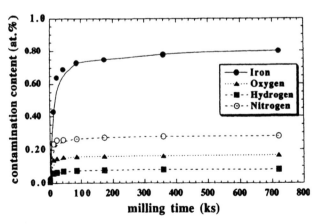

Fig. 4.7. Variation of the contamination contents in the milled Co$_{75}$Ti$_{25}$ powder as a function of the MA time

vial reached to about 320 K (almost every 1.8 ks of continuous milling) and then resumed when the temperature decreased to 300 K. In order to estimate the effect of the temperature increase on the structure of the powder during milling, the samples milled for 11 ks (amorphous phase of $Co_{75}Ti_{25}$ alloy) and 86 ks (bcc-Co_3Ti) were heated separately in the DTA to 700 K (well above the measured vial temperature). The XRD patterns of these samples prove that no phase transformation takes place even at this relatively high temperature, having the same XRD patterns that displayed in Fig. 4.1b and c. It is worth noting that the bcc-Co_3Ti phase does not change to any other phases, even after heating to higher temperature (1300 K) as was confirmed by the X-ray diffraction analysis. However, the amorphous $Co_{75}Ti_{25}$ phase transforms to the crystalline fcc-Co_3Ti phase upon heating to 880 K, as shown by the DTA measurements (see Fig. 4.6). We can then conclude that the temperature which recorded during the ball-milling process is much lower than the required temperature which can cause a structural changes in the milled powder. The same results have been obtained for the samples which milled independently in the second and third milling runs.

The cyclic crystalline-to-amorphous phase transformations occurs during ball milling of elemental $Co_{75}Ti_{25}$ powders. Such cyclic phase transformations are not caused by the contamination of the powder and/or the increase in the vial temperature during the milling. After a very short milling time (11 ks) an amorphous phase of $Co_{75}Ti_{25}$ is formed. This amorphous phase does not withstand against the impact and shear forces that are generated by the milling media (balls) and transforms completely into a new metastable bcc-Co_3Ti phase upon milling for 86 ks. This bcc-Co_3Ti phase is thermally stable and does not transform to any other phase(s) upon heating to 1300 K. It returns to the same amorphous phase of $Co_{75}Ti_{25}$ upon milling for 360 ks. This phase transformation is attributed to the accumulation of several lattice imperfections such as point and lattice defects which raise the free energy from the stable phase to a less stable phase (amorphous). The amorphous $Co_{75}Ti_{25}$ phase transforms to the fcc-Co_3Ti phase upon heating up to 874 K.

4.3 Formation of Amorphous and Nanocrystalline Ni-W Alloys by Electrodeposition and Their Mechanical Properties

4.3.1 Electrodeposition – A Method for the Production of the Amorphous Materials

Electrodeposition is a superior technique for producing amorphous and nanocrystalline materials in bulk form or as coatings with no post-processing requirements [17]. For engineering applications, amorphous and nanocrystalline Ni-W alloys produced by electrodeposition might have some excellent properties such as high hardness, high corrosion resistance and high thermal stability. However, this process for the Ni-W alloys has not yet been well developed

[18–20]. In our previous study, we developed an aqueous plating bath for the Ni-W electrodeposition that yields amorphous alloys of fairly high tungsten content [21,22]. In the present study, the plating conditions for producing the amorphous and nanocrystalline electrodeposited Ni-W alloys have been studied in details. In addition, nanocrystallization behavior of the amorphous Ni-W alloys has been studied, and grain size dependence of the strength of nanocrystallized Ni-W alloys having grain sizes of a few to about 250 nm in diameter has been investigated by means of hardness measurements.

4.3.2 Preparation of Ni-W Alloys and Technique Used for Studies

The plating bath compositions and conditions selected for this study are shown in Table 4.2. Trisodium citrate and ammonium chloride were introduced as complexing agents to form complexes with both Ni and W in the plating bath solution. To improve conductivity, sodium bromide was used. Electrodeposition was done on Cu substrates prepared by electropolishing, and a high-purity platinum sheet was used as the anode. The plating cells, 600 ml beakers, each containing 500 ml of the solution were supported in a thermostat to maintain the desired bath temperature. A fresh plating bath was made for each experiment using analytical reagent grade chemicals and deionized water.

Structural analysis of the electrodeposited alloys was performed by means of X-ray diffraction using Cu-Kα radiation (45 kV–25 mA), small angle X-ray scattering (SAXS, Cu-Kα, 40 kV–200 mA) and high resolution electron microscopy (HR-TEM, JEM-2010, 200 kV). Elemental concentrations of the electrodeposits were analyzed by energy-dispersive X-ray spectroscopy (EDS) in a scanning electron microscope. The samples thus prepared were annealed at various temperatures in air or vacuum of about 10^{-3} Pa. Vickers

Table 4.2. Plating Bath Compositions and Conditions for Ni-W electrodeposition

Nickel Sulfate, $NiSO_4\ 6H_2O$	0.06 mol/l
Trisodium Citrate, $Na_3C_6H_5O_7\ 2H_2O$	0.3 – 0.5 mol/l
Sodium Tungstate, $Na_2WO_4\ 2H_2O$	0.14 mol/l
Ammonium Chloride, NH_4Cl	0.0 – 0.5 mol/l
Sodium Bromide, NaBr	0.15 mol/l
Bath Temperature	333 – 363 K
PH	8.5 – 9.2
Current Density	20 A/dm^2

microhardness was measured by using the as-electrodeposited and the annealed samples on Cu substrates with a 0.02 kg load and a loading time of 15 s in cross section.

4.3.2.1 Formation of Ni-W Alloys by Electrodeposition. The tungsten content of the electrodeposited Ni-W alloys is strongly influenced by the ammonium chloride concentration in the solutions. As shown in Fig. 4.8, when the citric acid concentration was fixed between 0.3 and 0.5 mol/l at a bath temperature of 363 K, the tungsten content of the deposited alloys and their deposition rate increased sharply with increasing the ammonium chloride concentration. The tungsten content is also strongly influenced by the plating bath temperature. Figure 4.9 shows X-ray diffraction patterns of the Ni-W electrodeposited alloys for various plating bath temperatures. The tungsten content increased with increasing bath temperature, X-ray diffraction peaks of the deposited alloys broadened with increasing tungsten content and an amorphous pattern appeared at a tungsten content of more than about 20 at.%.

Figure 4.10 shows DTA measurements of the as-electrodeposited Ni-W alloys at a heating rate of 0.33 K/s. With the Ni-W alloy electrodeposited at the bath temperature of 333 K, no distinct peaks were observed. When the bath temperature is 348 K and above, the amorphous X-ray diffraction pattern appeared and the crystallization of these amorphous Ni-W alloys took place in two steps. The first-step crystallization at the starting temperature

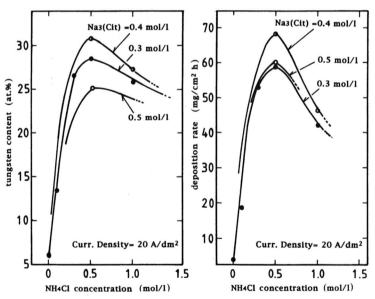

Fig. 4.8. Tungsten content of Ni-W electrodeposits and their deposition rate as a function of NH_4Cl concentration in the solutions

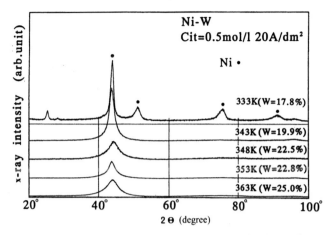

Fig. 4.9. X-ray diffraction patterns of Ni-W electrodeposited alloys for various plating bath temperatures

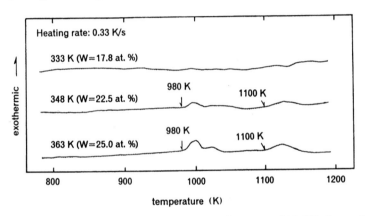

Fig. 4.10. DTA thermograms for crystallizarion of Ni-W electrodeposited alloys for various plating bath temperatures at a heating rate of 0.33 K/s

of about 980 K has been confirmed by X-ray analysis to be due to the precipitation of fcc-Ni solid solution. The second-step crystallization takes place at a temperature range of 1100–1150 K. X-ray analysis has suggested that Ni_4W intermetallic compound precipitates at this step accompanied by a formation of some kinds of oxides.

Deposition rate vs. plating bath temperature relationships for the electrodeposited Ni-W alloys are shown in Table 4.3. The deposition rate increases with increasing the plating bath temperature up to about 348 K and then decreases with increasing the bath temperature. It may be noted that the as-electrodeposited Ni-22.5 at.% W alloys at the plating bath temperature of 348 K are quite ductile: they can bend through an angle of 180° without breaking. Figure 4.11 shows SEM micrographs of the as-electrodeposited Ni-

Table 4.3. Deposition rate vs. plating bath temperature relationships for the Ni-W electrodeposited alloys (Na$_3$ (Citric Acid)= 0.5 mol/l in the solutions)

Plating Bath Temp. (K)	W content (at %)	Deposition rate (mg/cm^2 h)	Mechanical Properties
333	17.8	49.2	brittle
343	19.9	65.0	brittle
348	22.5	68.8	ductile
353	22.8	66.7	brittle
363	25.0	55.1	brittle

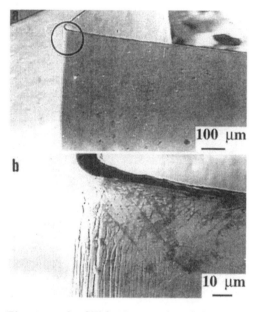

Fig. 4.11 a,b. SEM micrographs of the as-electrodeposited Ni-22.5 at.% W alloys after bending through an angle of 180°

W alloys after bending through an angle of 180°. They deform plastically and extremely inhomogeneously: shear bands form on the bending edge showing the typical feature of the ductile metallic glasses.

4.3.2.2. Nanocrystallization of the Ni-25.0 at.% W Alloys and Their Hardness Measurements. The specimen of the Ni-25.0 at.% W alloy with an amorphous structure were electrodeposited for examining thermal stability of their structure and mechanical properties by means of hardness measurements. Figure 4.12 shows X-ray diffraction patterns of the Ni-25.0 at.% W

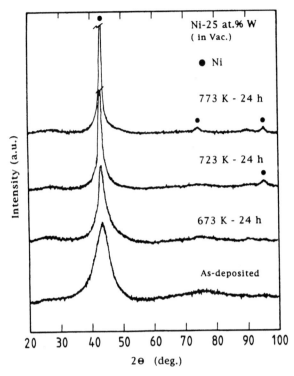

Fig. 4.12. X-ray diffraction patterns of Ni-25.0 at.% W amorphous electrodeposited alloys annealed at various temperatures for 24 h in vacuum

alloy annealed at various temperatures for 24 h in vacuum. On annealing at 673 K the amorphous structure was maintained, while the X-ray diffraction peak became sharper as compared with the as-deposited one. At higher annealing temperatures of 723 K and above, crystallization occurred and the diffraction lines of fcc-Ni phase appeared. In this case, a strong (111) fiber texture may be inferred from relative intensities of the diffraction peaks.

Figure 4.13 shows the Guinier plots of SAXS intensities obtained from the as-deposited specimen and the annealed one at 623 K for 24 h. For the as-deposited one, straight line relationship is observed. The slope of this line yields a value of about 2 nm in diameter for the size of domains of SAXS in the amorphous Ni-W alloy. This result suggests that there is a medium-range order in the amorphous phase. On annealing at 623 K, the slope of the Guinier plot increased continuously toward the observed smallest S-values, giving the evidence that the size of the domains is distributed on a broad scale. This may be due to the formation of nuclei for the nanocrystallization of the Ni-W alloy. The size distribution of the domains obtained by the Fankuchen method [23] is as follows; 1.7 nm (81 vol.%), 4.2 nm (15 vol.%) and others (4 vol.%) in diameter.

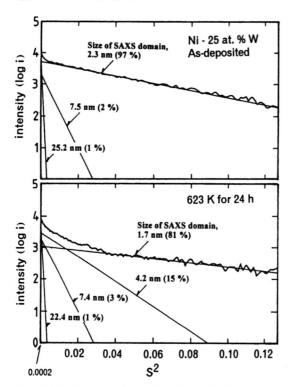

Fig. 4.13. Guinier plots of small angle X-ray spectra of Ni-W amorphous electrodeposited alloys. S is a scattering vector ($S=4\pi \sin\theta/\lambda$)

Figure 4.14 shows the Vickers microhardness vs. $d^{-0.5}$ (d is the mean grain diameter) relationship for Ni-W alloy after annealing at various temperatures in air or vacuum. Average grain sizes in the Ni-25.0 at.% W alloys were obtained by applying the Scherrer formula to the diffraction lines of fcc-Ni (111) and the broad maximum of the amorphous phase, and also by direct observations from TEM or SEM micrographs. For comparison, the relationships for electrodeposited Ni [24] and conventional pure Ni extrapolated from the relationship at larger grain sizes [25] are also shown. In the Ni-W alloy, the hardness increases with decreasing the grain size to about 10 nm. The Hall-Petch slope of $0.8 \text{ MPa m}^{1/2}$ in the Ni-W alloy is slightly higher, but comparable to those of the electrodeposited Ni and the conventional one. When the grain size is less than about 10 nm, the hardness decreases with decreasing grain size.

Figure 4.15 shows the TEM images and the selected area diffraction patterns in the Ni-25.0 at.% W alloy annealed at 723 K and 873 K for 24 h in vacuum. In Fig. 4.15a, nanocrystalline structure having grain sizes between 5 and 8 nm is observed. The selected area diffraction pattern reveals the fcc-lattice Debye rings indicating that the ultrafine grains become randomly oriented. Noticeably, image contrast of the interface of individual grains is not

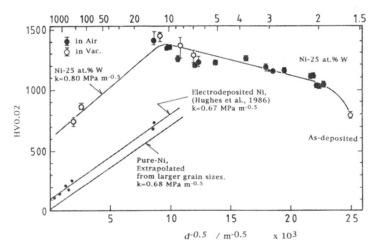

Fig. 4.14. Vickers microhardness as a function of $d^{-0.5}$ (d is the mean grain diameter) relationship for the Ni-25.0 at.% W alloy after annealing at various temperatures in air or vacuum

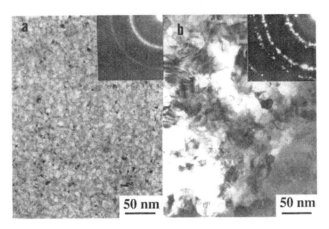

Fig. 4.15. TEM images and selected area diffraction patterns in the Ni-25.0 at.% W alloys annealed at 723 K (Fig. 4.8a) and for 873 K (Fig. 4.8b) for 24 h in vacuum

clearly visible. There is a possibility that such broad intercrystalline regions might have an irregular structure such as an amorphous. On annealing at 873 K, grain sizes increased to about 15 nm and above. Figure 4.16 shows the high resolution fcc-(111) lattice image of the Ni-W alloy annealed at 723 K for 24 h in vacuum. In the intercrystalline regions that are about 1 to 2 nm in widths, distorted lattice images are observed. As shown in the framed part of Fig. 4.16, the distorted lattice images were observed in the intercrystalline regions having components of grain boundary and grain boundary triple junction. Especially, highly distorted lattice image is observed in the region of triple junction (i.e., intersection line of three or more adjoining grains). When

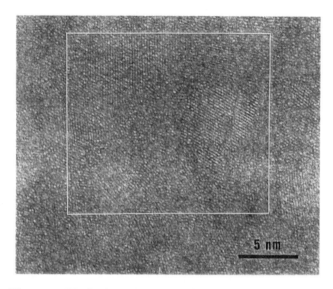

Fig. 4.16. The high resolution fcc- (111) lattice image showing 5 to 8 nm crystallites of the Ni-25.0 at.% W alloy annealed at 723 K for 24 h in vacuum

the defocusing value was changed from −48 nm to +16 nm, the straight line lattice image could not be observed in these intercrystalline regions. Similar features of HR-TEM observations have been reported in Ti-Mo alloys prepared by mechanical alloying and Ni_3Al by magnetron sputtering with grain sizes of less than 10 nm [26,27].

4.3.3 Brittleness of the As-electrodeposited Ni-W Alloys

It is well known that electrodeposited Ni-P amorphous alloys are severely brittle whereas melt-quenched Ni-P amorphous alloys exhibit high ductility with completely bending. Suzuki et al have suggested that the brittleness of the electrodeposited amorphous alloys may be caused by inclusion of hydrogen, water and anions in an electroplating solutions during depositing process [28].

As indicated in Table 4.3, the amorphous electrodeposited Ni-22.5 at.% W alloys having high ductility with completely bending was obtained at the plating bath temperature of 348 K. At this temperature the deposition rate (i.e., the current efficiency for the electrodeposition) of the Ni-W alloys was attained to the maximum value, indicating that the amount of codeposited hydrogen on the cathode during the electrodeposition should be a minimum. In order to clarify the effect of the codeposited hydrogen on the brittleness, the restoration behavior of the brittleness of the amorphous Ni-25.0 at.% W alloy produced at the plating bath temperature of 363 K was examined by measuring the fracture strain of the samples after annealing at various temperatures shown in Fig. 4.17. The fracture strain was measured by a simple bend test at room temperature. As shown in this figure, the fracture strains

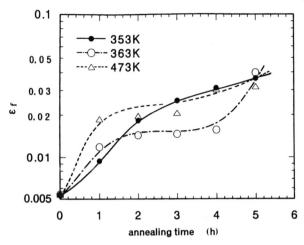

Fig. 4.17. Fracture strain vs. annealing time relationships at 353 K, 363 K and 473 K for the as-electrodeposited amorphous Ni-25.0 at.% W alloys

rise at the initial periods of annealing at 353 K, 363 K and 473 K respectively and then gradually rise with increasing annealing times.

These results suggest that the brittleness of the electrodeposited Ni-W alloys may be caused mainly by inclusion of the codeposited hydrogen. The effect of other factors on the brittleness such as an internal residual stress and inclusion of anions from the plating solutions are matters of future research.

4.3.4 Hardness of the Nanocrystalline Ni-W Alloys

Grain boundary structure of conventional coarse grained metals and alloys is well represented by the coincidence site lattice model (CSL-model) having periodic tilt boundary structures indexed by values [29]. In the case of the Ni-W alloys shown in Fig. 4.14, the Hall-Petch strengthening was observed for the hardness extending to a finest grain size of about 10 nm. Therefore, when the grain size is above 10 nm, the grain boundary structures may be essentially the same as those of the coarse grained materials.

When the grain size is less than about 10 nm, decrease of the hardness was observed. Ovid'ko has proposed that quasi-periodic tilt grain boundaries having a structure disordered in a gas-like manner exist in nanocrystalline materials [30]. As shown in Fig. 4.16, grain boundary thickness of about 1 to 2 nm evaluated from the HR-TEM observation may be considerably thicker than that of the coarse grained materials (almost less than about 1 nm). As a result, the volume fraction of the grain boundary triple junction having the highly distorted lattice image should increase remarkably with decreasing the grain size. Palumbo et al. have proposed to evaluate grain size dependence of volume fractions associated with both the grain boundary and the triple junction [31,32]. The total intercrystalline fraction (V_f^{ic}) and the grain boundary fraction (V_f^{gb}) are derived from the following equations,

$$V_f^{ic} = 1 - \left[\frac{(d-D)}{d}\right]^3, \tag{4.1}$$

$$V_f^{gb} = \frac{3D(d-D)^2}{d^3}, \tag{4.2}$$

where d and D are the grain diameter and the grain boundary thickness, respectively. The fraction associated with the triple junction (V_f^{tj}) is given by

$$V_f^{tj} = V_f^{ic} - V_f^{gb}. \tag{4.3}$$

Figure 4.18 shows the calculated volume fractions for V_f^{ic}, V_f^{gb} and V_f^{tj} as a function of $d^{-0.5}$ where D is changed over a range from 0.5 to 2.0 nm. Assuming that the formation of size distribution of the SAXS-domains in Fig. 4.13 is due to the formation of nanocrystallization nuclei, the V_f^{ic} can be obtained by equating with the fraction of the SAXS-domains having sizes of less than 2 nm. The data shown by filled circles in Fig. 4.18a are the V_f^{ic} estimated from the SAXS spectra of the Ni-W alloys as-electrodeposited and annealed at 623 K for 24 h.

As shown in this figure, these data are in good agreement with the values of theoretical expectations having grain boundary thickness between 1.5 and 2.0 nm. The V_{gb}^f in Fig. 4.18b increases to maximum values of about 45% and then decreases with decreasing the grain size, while the V_f^{tj} in Fig. 4.18c continues to increase. When the grain boundary thickness is 1.5 nm and above, the marked increase of the V_f^{tj} is observed in the range of grain sizes of less than about 10 nm.

Figure 4.19 shows the Vickers microhardness vs. $d^{-0.5}$ relationship for the Ni- 25.0 at.% W alloy. In order to explain the decrease in hardness where the grain size is less than about 10 nm, we give a rough estimate of the grain size dependence of the hardness (H_{cal}) by following equation,

$$H_{cal} = V_f^{tj} H_{tj} + (1 - V_f^{tj}) H_{HP}, \tag{4.4}$$

where V_f^{tj} and H_{tj} are the volume fraction and the hardness of the triple junction, respectively, and H_{HP} is a hardness obtained by the Hall-Petch relationship for the Ni-W alloy. On the basis of the calculated volume fractions shown in Fig. 4.12, we assume that the hardness of the triple junction (H_{tj}) is equal to that of the as-electrodeposited amorphous Ni-W alloy (H_v = 770). The H_{cal} vs. $d^{-0.5}$ relationships where the grain boundary thickness is changed in a range of between 0.5 and 2.0 nm are also shown in this figure. The decreasing tendency of the hardness under conditions where $D = 1.5$ and 2.0 nm can be well expressed by (4.4). Therefore, the significant increase of the triple junction fraction may be responsible for decreasing the hardness.

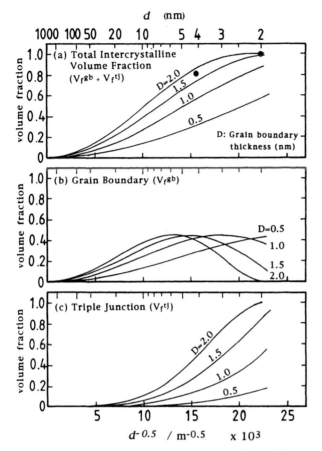

Fig. 4.18a–c. Calculated volume fractions for total intercrystalline volume fraction (V_f^{ic}), grain boundary volume fraction (V_f^{gb}) and triple junction volume fraction (V_f^{tj}) as a function of $d^{-0.5}$ where D is changed over from 0.5 to 2.0 nm. The data shown by filled circles are V_f^{ic} estimated from the SAXS spectra of the Ni-25.0 at.% W alloys

As mentioned above, when the grain size is less than about 10 nm, the grain boundary thickness evaluated from the HR-TEM observation may be considerably thicker than that of the coarse grained materials. Fujita has speculated atomic structure of ultrafine crystallites in a nanometer scale by using a volume free energy difference and a surface energy of an atom cluster [33]. Interface region of the nanocrystallites having a structure of non-periodic atomic array gradually expands into the center region when the size of crystallites decreases below a critical one. Swygenhoven et al. [34] have also calculated the influence of grain size on the mechanical properties of nanostructured Ni with the grain sizes between 3 and 10 nm by a molecular dynamics computer simulation, and have proposed terms of grain boundary viscosity giving rise to a viscoelastic behavior with the grain sizes below 5 nm. These results support that the grain boundary thickness may increase with decreasing the grain size, and the hardness of the expanded grain boundary region may be lower than that of the inner grain region. Finally, the nanocrystalline structure can be transformed into the amorphous state under the condition where the grain boundary thickness is equal to the grain size. In order to clarify such a phase transformation and a softening mechanism, further experiments will be carried out, especially HR-TEM observations as a function of the grain size. The tungsten content of the Ni-W electrodeposits is strongly influenced by ammonium chloride concentration in the plating bath and is also influenced by the plating bath temperature. The X-ray diffraction peaks of the deposited alloys broadened with increasing tungsten content and an amorphous pattern appeared at a tungsten content of more than about 20 at.%. Noticeably, the amorphous electrodeposited Ni-22.5 at.% W alloys having high ductility with completely bending was obtained at the plating bath temperature of 348 K. It is suggested that the ductility of the electrodeposited Ni-W alloys may be strongly influenced by inclusion of the codeposited hydrogen during depositing process. By annealing the Ni-25.0 at.% W alloy, a significant increase of the hardness occurs,

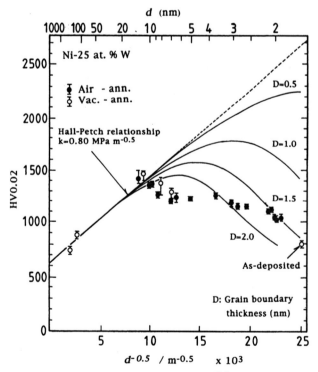

Fig. 4.19. Vickers micro-hardness vs $d^{-0.5}$ relationship of the Ni-25.0 at.% W alloy. The H_{cal} vs. $d^{-0.5}$ relationships where the grain boundary thickness is changed in a range between 0.5 and 2.0 nm are also shown

and the Hall-Petch strengthening was observed for the hardness extending to a finest grain size of about 10 nm. When the grain size is less than about 10 nm, the hardness decreases with decreasing the grain size. In the case of the 723 K-annealed specimen having grain sizes about between 5 and 8 nm, the distorted lattice image with about 1 to 2 nm thickness in grain boundary regions was observed by HR-TEM. Especially, highly distorted lattice image was observed in the regions of grain boundary triple junction. Such a thickness of the grain boundary may be considerably thicker than that of the coarse grained materials. As a result, the volume fraction of the grain boundary triple junction having highly disordered structure should increase remarkably with decreasing the grain size. Therefore, this decrease in hardness may be due to the significant increase of the intercrystalline volume fraction, especially the fraction associated with the triple junction.

4.4 Formation of Ti-Based Amorphous Alloys by Sputtering and Their Physical Properties

4.4.1 Sputtering Technique

Application of materials in new engineering fields needs new functioning alloys. New alloys have been developed by using new metastable materials produced by passing from the gas state directly to the solid state. Making amorphous alloys from the crystalline structures gives alloys high corrosion resistance because amorphous alloys have no defects of precipitates and crystalline boundaries that are initiation sites for corrosion [35]. Naka et al. [36–38] reported that amorphous stainless iron, cobalt and nickel alloys show high corrosion resistance. Titanium is also another candidate material for corrosion resistant alloys. Amorphous titanium alloys that contain P as a glass-forming element and were produced from liquid state exhibited high corrosion resistance [39]. Amorphous alloys have been prepared by rapid quenching of molten alloys on a highly rotating copper wheel to obtain high cooling rates. The high cooling rate prevents the nucleation of crystals from liquids. Vapor quenching also provides an easy process for making amorphous alloys which can contact various substrates [40]. This paper discusses titanium base alloys containing alloying elements of B, Si, C or Al prepared by a sputtering process and also describes physical properties such as the structure, microhardness, and crystallization temperature of the alloys.

4.4.2 Samples Preparation and Description of the Analytical Equipment Used

RF magnetron sputtering was used to prepare amorphous alloys in a low pressure argon gas at 6.65 MPa. The schematic configuration of the sputtering apparatus is presented in Fig. 4.20. Targets used were 100 mm in diameter and

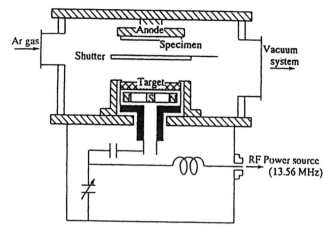

Fig. 4.20. Magnetron sputtering apparatus

5 mm thick and were composed of Ti and B, Ti and Si or Ti and Al. The substrates of aluminum were water cooled. The main sputtering conditions were a sputtering power and time of 600 W and 14.4 ks, respectively. The film thickness of the alloys prepared was 20 μm. The structure of the sputtered films was investigated by X-ray diffractometry with CuKα. The mechanical properties and thermal stability of the films were measured by using a Knoop microhardness tester and a differential scanning calorimeter, respectively. The thermal stability of the films was also investigated by a differential thermal calorimeter with a heating rate of 0.33 K/s.

4.4.3 Structure and Mechanical Properties of Sputtered Alloys

Figure 4.21 shows the X-ray diffraction patterns of sputtered Ti-B alloys. The structures of alloys of No. 1 (0 at.% B) and No. 2 (2.9 at.% B) are the hcp crystalline phase. Though the crystalline titanium does not dissolve any content of B in the equilibrium state, the solid solution supersaturated with about 8 at.% B is formed by sputtering. The Ti-B alloys containing B content from 8 to 51 at.% which show only a few diffuse diffraction peaks are amorphous. By adding 51 at.% or more B the Ti-B alloys become crystalline phases composed of Ti_3B_4 and TiB_2. The addition of B to Ti makes Ti-B amorphous alloys in a wide range of B content. It is known that alloys that show deep suppression of the liquid temperature in eutectic reactions easily become amorphous. The melting points of Ti-B alloys are definitely suppressed by adding B. This may imply that the amorphous structures of Ti-B alloys are also stabilized by adding B to Ti. Amorphous Ti base alloys containing P [35] were prepared by a liquid quenching technique. The liquid quenching process restricts the composition range of amorphous alloys and the size of the alloys in ribbon type tape, the vapor quenching process provides a wide composition at range of amorphous alloys and is also useful

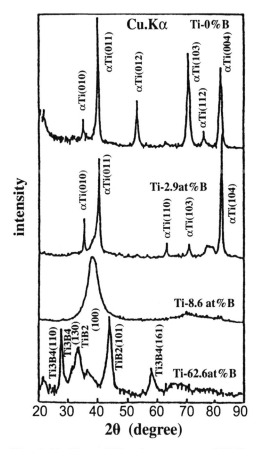

Fig. 4.21. X-ray diffraction patterns of Ti-B sputtered alloys

for applying coated films on various substrates. Figure 4.22 shows the X-ray diffraction patterns of Ti-Si sputtered alloys containing Si content from 0 to 100 at.% Si. The sharp X-ray diffraction peaks of the crystalline hcp phase are indicated in pure Ti, Ti-6.4 at.% Si and 12.0 at.% Si alloys. The silicon content of 12 at.% in a solid solution formed by sputtering is two times higher than that at equilibrium. Because Ti-11.5 at.% Si and Ti-Si alloys containing 30 at.% or more Si have the diffuse X-ray diffraction patterns of amorphous phases, the sputtered Ti-Si alloys containing 12 at.% Si or more are amorphous. Ti-Si alloys containing around 80 at.% Si have two first X-ray peaks at 30 and 40 degrees, and are composed of two amorphous phases of Ti rich and Si rich phase. Compared with the glass formation range of Ti-B alloys sputtered, that of Ti-Si sputtered alloys is wider, consisting of Si content from 12 to 100 at.% Si.

TiC sputtered alloy is formed at a C content of 35 at.% or less, which corresponds to the lowest C content of TiC at the equilibrium state. In other words, TiC with a C content, that deviates from the stoichiometric carbon

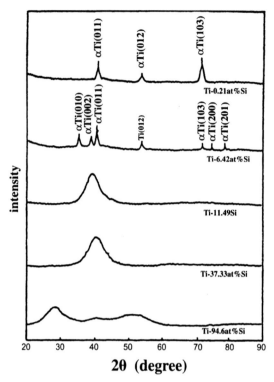

Fig. 4.22. X-ray diffraction pattern of Ti-Si sputtered alloys

content in TiC is formed by sputtering. Ti$_3$Al in the sputtered alloys is formed at an Al content of 15 at.%, though it is formed at an Al content of 20 at.% or more at room temperature in equilibrium state. The admixture of alloying elements in Ti based binary sputtered alloys forms amorphous alloys with random atomic structures as shown in Fig. 4.23. For instance, amorphous alloys are formed at B content from 8 to 50 at.%, Si content of 12 at,% or more, or Al content from 41 to 65 at.%, respectively. Except for Ti-C system containing strong carbide former, the admixture of elements of a certain content causes the formation of amorphous structure with the random atomic locations. The mixing entropy terms reduce the free energy of the amorphous phase, and accounts for the formation of the amorphous structure. The compositional dependence of microhardness for the sputtered alloys is shown in Fig. 4.24. The microhardness of Ti-C alloys, in which the atomic interaction is strong, increases extremely with C content. Ti-B or Ti-Si yields the same increase in the microhardness with increase in alloying content up to 50 at.% B or 40 at.% Si. In Ti-Si alloys, the microhardness of the Ti alloys decreases when the silicon content increases, corresponding to composition the content of Ti$_5$Si$_3$ compounds. This suggests that the alloys whose composition correspond to a compound, such as TiB or Ti$_5$Si$_3$ in Ti-B or Ti-Si alloys

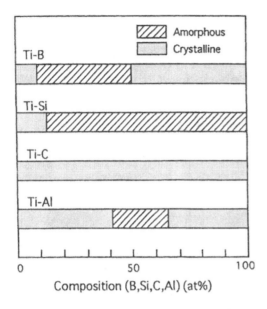

Fig. 4.23. Structure and composition of sputtered Ti alloys.

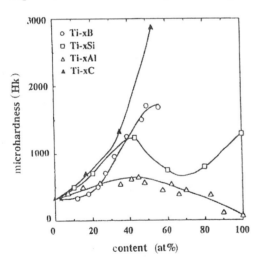

Fig. 4.24. Compositional dependence of microhardness of Ti alloys

yield the high hardness. In other words, amorphous alloys also have the local atomic structure similar to the corresponding crystalline compounds. Ti-Al alloys also show maximum microhardness around on Al content of 50 at.%, corresponding to an AlTi compound.

4.4.4 Amorphous to Crystalline Phase Transition

Although the glass formation compositional range of Ti base alloys differs for each system, the amorphous alloy is formed by adding an alloying element. Amorphous alloys have the superior properties such as mechanical properties, alloys in the metastable state will transform to the crystalline phases in a stable state at a certain temperature.

Figure 4.25 shows the compositional dependence of crystallization temperatures of amorphous Ti-B, Ti-Si and Ti-Al alloys. Amorphous Ti-B alloys posses the higher crystallization temperatures than the other Ti base alloys except for the Ti rich Si containing alloys. In particular, amorphous Ti -50 at.% B alloy possesses the highest crystallization temperature of about 1000 K. Amorphous Ti-Si alloys containing up to 50 at.% Si crystallize from 800 to 850 K. A further increase in the Si content up to 67 at.% sharply lowers the crystallization temperature. Amorphous alloys whose compositions are the same as intermetallics such as TiB, Ti_5Si_3 or Ti_3Si, Ti_3Si and $TiSi_2$ have higher crystallization temperature than other amorphous alloys of the composition apart from the intermetallics. The amorphous alloys with the compositions of the intermetallics are expected to crystallize polymorphously.

Amorphous alloys are likely to have structures similar to the corresponding liquid alloys. Then the crystallization temperature should be normalized by the melting temperatures. Fig. 4.26 shows the crystallization temperature of amorphous Ti-Si alloys which increase with the melting temperature. Amorphous Ti-Si alloys containing 67 at.% Si or more apparently yield the increase in the crystallization temperature with the Si content. These alloys may crystallize with a behavior different from the other alloys.

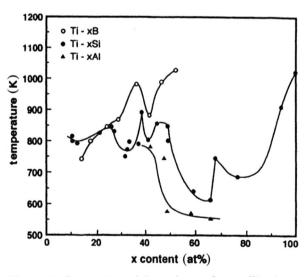

Fig. 4.25. Compositional dependence of crystallization temperatures of amorphous Ti-B, Ti-Si and Ti-Al alloys

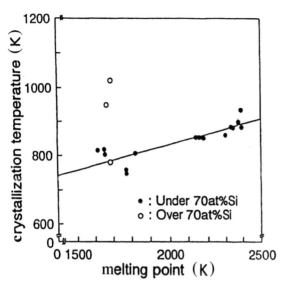

Fig. 4.26. Relationship between crystallization temperature of amorphous Ti-Si alloys and melting points for Ti-Si alloys with equivalent compositions

The metastable Ti based alloys containing B, Si, C or Al were prepared by sputtering in a low pressure argon. The amorphous phases are formed by mixing certain amounts of alloying elements, except for the Ti-C system which contains a strong carbide former carbon. The entropy terms resulting from adding Ti and elements reduce the free energy of the random structure, and account for the formation of amorphous structures. Amorphous alloys yield the superior mechanical properties such as microhardness. Amorphous Ti-base alloys that have compositions of intermetallics show higher hardness and higher crystallization temperatures than the other alloys. This suggests that amorphous alloys possess local atomic structure similar to the corresponding crystalline compounds of TiB, Ti_5Si_3 and TiAl.

4.5 Hydrogen Evolution Characteristics of Ni-Mo Alloy Electrodes Prepared by Mechanical Milling and Sputter Deposition

4.5.1 CO_2 Recycling Problem

In order to reduce CO_2 emission which induces global warming, we have been proposing CO_2 recycling [41] which requires a large scale production of hydrogen by the electrolysis of seawater. In this recycling, the electric power is generated by the solar cell operation at deserts and transmitted to the nearest coasts. Using this electricity H_2 can be generated by the electrolysis of seawater. On the other hand, CO_2 evolved at combustion plants is recovered, liquefied and transported to the coasts by tankers. The CO_2 is converted into CH_4 by the reaction with H_2. The produced CH_4 is liquefied and transported to energy consumers by tankers and used as a fuel. It is, therefore, necessary to develop highly active and stable electrocatalysts for hydrogen evolution reaction (HER) without using noble metals.

In the past decade, many papers have dealt with ways of increasing the effectiveness of cathodes used for the HER in alkaline solutions [42,43]. Many transition metal alloys have been characterized as hydrogen electrodes. Among them, nickel and nickel-based alloys [44,45], particularly Ni-Mo alloys [46–51], have a high catalytic activity. However, their activity and stability are not sufficient. They are usually prepared by electrolytic codeposition [45,46], dip coating [47], flame [49] or plasma [50] spraying or thermal decomposition with subsequent reduction to metal alloys [51].

The homogeneous solid solution alloys, including amorphous alloys, are suitable for preparation of materials with high catalytic activity, since it is possible to modify the electronic state of active elements as well as synergistic effects by alloying with various elements. There have been many reports to show the extremely high electrocatalytic activity and stability for some reactions such as electrolysis of sodium chloride solution [52] and seawater [53], oxygen evolution [54], and oxidation of hydrogen [55] and methanol [56]. In recent years, amorphization by high energy mechanical alloying (MA) of pure crystalline powders was reported [57,58]. It is interesting to note that nanocrystalline Ni-Mo powders prepared by MA were found to have high electrocatalytic activity for HER in 30% KOH at 70°C [48]. The sputtering technique is also quite useful for the preparation of homogeneous solid solution alloys and well-suited for coating application.

The present work aims to compare the electrocatalytic activity for HER of Ni-Mo alloy electrodes prepared by mechanical alloying with the electrodes made by sputter-deposition in a sodium hydroxide solution.

4.5.2 Experimental Procedure

4.5.2.1 Preparation of Electrodes

4.5.2.1.1. Mechanical Alloying. Mixtures of elemental powders of nickel (99.9%, 100 mm in diameter) and Mo (99%, 3 mm in diameter) were mechanically alloyed using a planetary ball mill (Fritsch P5) to give nominal composition of Ni-(20, 30, 40 and 50) at.% Mo. The ball to powder weight ratio was 10:1. Before introducing argon gas, the vial was evacuated using a diffusion pump up to 1×10^{-4} torr. The ball milling was performed in a cylindrical tool steel (SKD 11) vial in an argon atmosphere together with high-carbon chromium steel balls. The mill was rotated at 150 rpm. Mechanically alloyed powders were pressed at 3.9–8.6 tons/cm² to form an electrode and mounted in an epoxy resin. However, some powders were pressed with binder (polytetrafluoroethylene, PTFE, suspension) and then heated at 350°C for 4 h in argon, since it was difficult to consolidate the powders with higher Mo content or longer milling time without addition of the PTFE binder. The powder to PTFE weight ratio was 25:1.

4.5.2.1.2. Sputter Deposition. Ni-Mo alloys were prepared by a sputter-deposition method using D.C. magnetron technique described elsewhere [59,60]. Nickel plate (99.5%), 0.2 mm thick and glass plate, 1 mm thick were used as substrates. Before sputter-deposition, nickel substrates were polished by SiC paper up to # 1500, degreased in acetone and annealed at 600°C for 2 h in vacuum. They were chemically etched in 50% H_2SO_4+50% HNO_3 for 1–2 min at 50°C, followed by washing with a splash of distilled water. Glass plates were cleaned by immersion in a commercial detergent for aluminum metal at 75°C, followed by rinsing with distilled water. After the target and the substrates were installed in the sputtering machine the vacuum chamber was evacuated to about 5×10^{-7} torr. Sputtering was performed for 4 h at about 9×10^{-4} torr of argon gas.

Surface activation treatment was performed by leaching of only the sputter-deposited Ni-Mo alloys in 10 M NaOH at 80°C from 1 to 60 min. This treatment is considered to induce the dissolution of molybdenum producing effective catalysts with small nickel particles as active phase and large surface area.

4.5.2.2. Electrochemical Measurements. The HER on the mechanically alloyed and sputter-deposited Ni-Mo alloy electrodes was examined in 1M NaOH solution at 30°C. The solution was deaerated by bubbling a stream of oxygen-free nitrogen for more than 15 h. Before the measurement of cathodic polarization curves, cathodic pre-polarization at 1 A/m^2 for 10 min was carried out. Then, after the electrolyte was changed with fresh one, cathodic polarization of the electrodes was performed at a potential sweep rate of 20 mV/min. The reference electrode was a saturated calomel electrode, and the counter electrode was a platinum gauze of a large surface area. Correction for ohmic drop was made using a Hokuto IR Compensation Instrument.

4.5.2.3. Characterization of Electrodes. The structure of mechanically alloyed powders and sputter-deposits were identified by X-ray diffraction with $CuK\alpha$ radiation. The composition of the sputter-deposits was determined by an electron probe micro-analysis. The apparent grain size of the alloys was estimated from the full width at half maximum (FWHM) of the main diffraction peak according to Scherrer's equation. The surfaces of the electrodes were characterized by using Shimadzu ESCA 850 electron spectrometer with Mg $K\alpha$ excitation. The binding energies of the electrons were determined by a calibration method described elsewhere [61,62]. Quantitative analysis was performed by the method reported previously [63,64]. The composition of the surface region of the electrodes was estimated without distinguishing the metallic and oxidized states because some of the atoms in metallic state might have been oxidized during specimen transfer from the solution to the XPS chamber. The photo-ionization cross-section of Ni $2p_{3/2}$ and Mo $3d_{3/2}$ electrons relative to the O 1s electrons used were 2.32 [63], and 3.42 [65], respectively. The surface morphology of the electrodes was observed by SEM.

4.5.3 Mechanically Alloyed Ni-Mo Electrodes

4.5.3.1. Structural Change with Milling Time. Figure 4.27 shows the X-ray diffraction patterns of Ni-40Mo (at.%) mixtures mechanically alloyed for different periods of milling time. The reflections of fcc nickel and bcc molybdenum are clearly seen in XRD patterns of the sample after mechanical alloying (MA) for 2 h. After MA for 12 h, the relative peak intensity decreases, while the FWHM of each reflections increases. As the milling proceeds, the Bragg reflections of nickel disappear while the reflections of molybdenum are still observed on the specimen even after 192 h of MA. The shift of nickel reflections toward the low angles together with the reduction of molybdenum

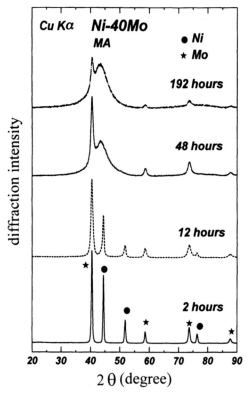

Fig. 4.27. X-ray diffraction patterns for Ni-40Mo alloys ball-milled for different periods

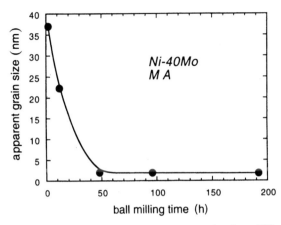

Fig. 4.28. The change in apparent grain size of Ni crystallites with milling time

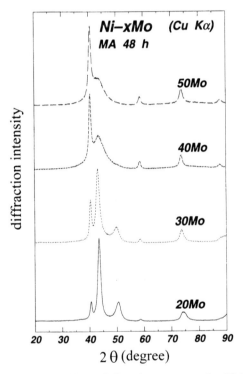

Fig. 4.29. X-ray diffraction patterns for Ni-Mo alloys ball-milled for 48 h

reflections indicates that molybdenum diffuses inside the nickel lattice and produces an expansion of the lattice. Figure 4.28 shows the apparent grain size of nickel crystallites as a function of milling time estimated from the full width at half maximum (FWHM) of the main diffraction peak according to Scherrer's equation. The apparent grain size decreases sharply with milling time and is the order of 2 nm for the specimen after MA for 48 h, which is only a little larger than the size of atom group supposed to exist in liquid metals. Figure 4.29 exhibits X-ray diffraction patterns of mechanically alloyed Ni-xMo powders of different composition after MA for 48 h. The amorphous phase tends to form with increasing molybdenum content but single-phase Ni-Mo alloy does not form after MA for 48 h.

4.5.3.2. HER Characteristics. In order to prepare the bulky electrode, the consolidation of cold die-pressing of the powders was carried out under the pressure of 3.9-8.6 tons/cm^2 at room temperature. However, the bulky electrodes were obtained only for the powders with short milling time up to 12 h. The consolidation of the powders with longer milling time was unsuccessful under the same condition probably due to the work hardening of the powders during MA. The pressed compacts were very crumbly. Therefore, PTFE binder was added to the alloy powders. Figure 4.30 shows cathodic polarization curves of Ni-40Mo alloy electrodes thus prepared with different milling time measured in deaerated 1M NaOH at 30°C. The polarization curves of arc-melted Ni-40Mo alloy and pure Ni plate (bulk) are shown for comparison. For pure nickel electrode, the Tafel slope of the low current density region is about 60 mV/dec and that of the high current region is about 140 mV/dec. However, for the ball milled electrodes, there is only a single line with Tafel slope of 140–170 mV/dec. As shown in the Fig. 4.31, strong dependence of the milling time on the hydrogen overpotential and the exchange current density for hydrogen evolution is observed. With milling proceeds, the hydrogen overpotential decreases, but after milling for 48 h it varies little with milling time. It is interesting to note MA electrodes ball-milled for

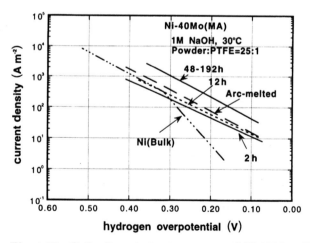

Fig. 4.30. Cathodic polarization curves of Ni-40 Mo alloy electrodes ball-milled for different periods and consolidate the powders by cold die-pressing with PTFE measured in 1 M NaOH at 30°C

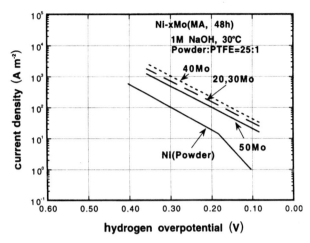

Fig. 4.31. Cathodic polarization curves obtained on Ni-Mo alloy electrodes ball-milled for 48 h and cold pressed the powder with PTFE measured in 1 M NaOH at 30°C

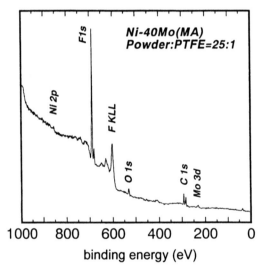

Fig. 4.32. XPS spectrum of ball-milled Ni-40Mo alloy containing PTFE obtained with a wide range scan of binding energy (0–1000 eV)

48–192 h show higher activity for HER compared with the arc-melted alloy electrode, although the hydrogen overpotential of the MA electrode is much higher than that of the ones reported for the Ni-Mo alloys [46–51].

Figure 4.31 shows cathodic polarization curves obtained on Ni and Ni-Mo alloy electrodes ball-milled for 48 h and cold die-pressing with PTFE. As may be seen, hydrogen overpotential changes with molybdenum content. The maximum activity for HER is obtained for Ni-40Mo alloy electrode although the activity is still very low. In addition, in the case of high current region

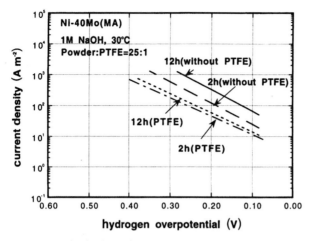

Fig. 4.33. Cathodic polarization curves obtained on ball-milled Ni-40Mo alloy electrodes with and without PTFE measured in 1 M NaOH at 30°C

more than $2-3 \times 10^3$ A/m^2, the current vibration was observed. The low catalytic activity and the current vibration may be attributed to the high ohmic resistance of PTFE contained in the electrode.

The surface analysis of these electrodes was conducted using XPS apparatus. Figure 4.32 shows the XPS spectrum for the wide energy scans (0–1000 eV) obtained from PTFE-containing Ni-40Mo electrode. The spectrum indicates the existence of nickel, molybdenum, fluorine, oxygen and carbon on the electrode. Obviously, the electrode surface is covered by a large amount of fluorine, which reduces electronic conductivity of the electrode and results in low activity. Therefore, the cathodic polarization curves obtained on the MA electrodes without PTFE were measured in 1M NaOH at 30°C. Figure 4.33 shows a comparison of the polarization curves for the Ni-40Mo electrodes ball-milled for 2 and 12 h and pressed with and without PTFE. The improvement of HER performance is observed for the electrodes without PTFE. The maximum activity is obtained for the electrode ball-milled for 12 h without PTFE, compared with Figs. 4.30, 4.31 and 4.33. It is, therefore, expected that the Ni-Mo alloy electrodes ball-milled for longer time should have high catalytic activity for HER. Unfortunately, as noted earlier, such electrodes were unsuccessfully prepared.

4.5.4 Sputter-deposited Ni-Mo Electrode

4.5.4.1. Characterization of Sputter-Deposition. Ni-(15-52) at.% Mo alloys have been prepared by the sputter deposition method. The average thickness was about 2 μm. Binary Ni-Mo alloys are amorphized in the wide composition range as shown in Fig. 4.34. The alloys containing more than 21 at.% molybdenum show the halo pattern typical of the amorphous structure. Compared with MA technique, shown in Fig. 4.29, it can be said that

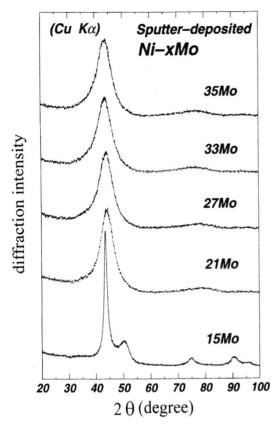

Fig. 4.34. X-ray diffraction patterns for sputter-deposited Ni-Mo alloys

the sputter deposition method has a great advantage as one of appropriate methods to prepare single phase amorphous alloys. The Ni-15Mo alloy, however, shows broad peaks corresponding to the 111, 200, 220, 311 and 222 reflections of the fcc nickel phase supersaturated with molybdenum.

These alloy deposits showed good adhesion to nickel substrates and no materials were detached from the electrodes during polarization although those on the glass substrates were peeled off during the evolution of hydrogen gas.

4.5.4.2. HER Characteristics. Figure 4.35 shows a comparison of potentiodynamic polarization curves for the Ni-Mo coated electrodes on nickel substrates in deaerated 1 M NaOH solution at 30°C with one for nickel metal plate. Two Tafel regions can be distinguished on the polarization curves. In the low and the high hydrogen overpotential ranges, the Tafel slopes varied between 30–50 mV/decade and 120–140 mV/decade, respectively. The Ni-Mo alloy electrodes clearly exhibit favorable performance relative to that of the nickel metal. Decreasing the molybdenum content significantly enhances the activity and the maximum activity is observed at 15 at.% Mo alloy. The hy-

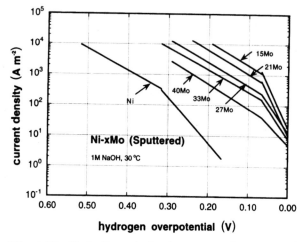

Fig. 4.35. Cathodic polarization curves for sputter-deposited Ni-Mo alloy electrodes in 1 M NaOH at 30°C

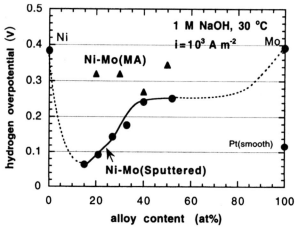

Fig. 4.36. Change in hydrogen overpotential at the current density of 10^3 A/m^2 with alloy Mo contents for sputter-deposited Ni-Mo alloys in 1 M NaOH at 30°C

drogen exchange current density for this alloy estimated by extrapolation of the Tafel line in the high hydrogen overpotential range to the zero overpotential is more than two orders of magnitude higher than that for bulk nickel.

Figure 4.36 shows relation between alloy molybdenum content and hydrogen overpotential for both Ni-Mo electrodes prepared by sputtering and MA together with the results of the parent metals and smooth platinum plate. It is clearly observed that the catalytic activity for HER of sputter-deposited alloy electrodes is remarkably higher than that of MA alloys, and furthermore, that of the alloy constituents. Therefore both nickel and molybdenum enhance the catalytic activity synergistically. The maximum activity is ob-

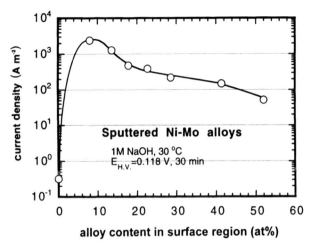

Fig. 4.37. Change in current density at the hydrogen overpotential of 0.118 V with Mo contents in surface region for sputter-deposited Ni-Mo in 1 M NaOH at 30°C

served at Ni-15Mo alloy electrode. Its hydrogen overpotential at 10^3 A/m^2 is about 67 mV which is lower than that of the catalytically active platinum electrode and is equivalent or lower than other Ni-Mo electrodes [46–51].

The activity of the electrode is strongly affected by their surface composition. Activating Ni-Mo alloy surface was evaluated by XPS. Figure 4.37 depicts the change in current density at the hydrogen overpotential of 0.118 V with alloy molybdenum content in surface region determined by XPS analysis. The maximum activity appears around 10 at.% Mo in the surface region.

The explanation of the high catalytic activity of Ni-Mo alloy has not yet been agreed upon. Conway et al. [66] attributed the catalytic activity of such alloys to nickel hydride formation which enhances the hydrogen evolution rate by a corrosion mechanism. Kita [67] and Miles and Thomason [68] studied the HER mechanism in relation to periodic variations of hydrogen overpotential with atomic number of the electrode metals. The overpotentials show minima for nickel, palladium and platinum which have d^8s^2, $d^{10}s^0$ and d^9s^1 electronic configurations, respectively. Large hydrogen overpotentials are observed for zinc, cadmium and mercury which are all of the $d^{10}s^2$ electronic configurations. It may be noted qualitatively from the electronic configurations of outer shells of the element that the activity in each long period increases first with successive addition of an electron to the d-orbital, reaches a maximum at nearly filled d-orbital, decreases quite sharply after its completion, with one or two electrons in the s-orbital, and then increases again with further addition of electrons to the p-orbital. The saturation magnetization for Ni-Mo alloys at zero Kelvin has been determined and the Bohr magnetron number referred to one atom of alloy, which corresponds to the number of d-band vacancy, has been calculated by Marian [69]. The d-band vacancy of nickel in Ni-Mo alloys decreases with the addition of molybdenum

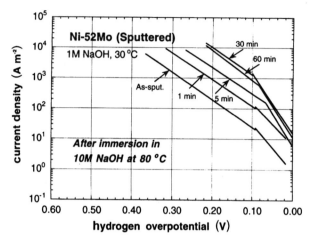

Fig. 4.38. Cathodic polarization curves for sputter-deposited Ni-52 Mo alloys in 1 M NaOH at 30°C after immersion in 10 M NaOH at 80°C for different time

and becomes nearly zero at about 11 at.% Mo. This is in good agreement with the fact that the optimum surface composition of HER activity for both alloys is about 10 at.% Mo. The metals having more unpaired electrons in the d-band interact strongly with electron donating metals or molecules and adsorb hydrogen strongly by electron pair formation between the hydrogen electron. The chemisorption of the hydrogen is necessary for HER but a slight decrease in the enthalpy for chemisorption may be effective to decrease the HER overpotential.

In order to obtain more active alloy electrodes, it is necessary to prepare the alloys containing molybdenum less than 15 at.%. However, the alloys thus prepared showed poor adhesion to nickel substrate. Therefore, an attempt was made to decrease molybdenum content in the surface region and to increase effective surface area by leaching treatment of the alloy with relatively high alloying content in hot concentrated caustic solutions. Figure 4.38 shows the cathodic polarization curves of Ni-52Mo alloys after immersion in 10 M NaOH at 80°C for different periods of time. The catalytic activity for HER increases with immersion time up to 30 min. Figure 4.39 illustrates the change in molybdenum content in the surface region for Ni-52Mo alloys with leaching treatment, followed by cathodic pre-polarization at 1 A/m² for 10 min (denoted as C.P.) and potentiostatic polarization at hydrogen overpotential of 0.118 V for 30 min in 1 M NaOH at 30°C. Increasing immersion time in 10 M NaOH solution results in the decrease in the molybdenum content. This leads to improvement of the catalytic activity shown in Fig. 4.38. Figure 4.40 shows AFM images before and after leaching in 10 M NaOH at 80°C, indicating clearly an increase in the surface roughness by the leaching treatment. Electrochemical impedance measurements also showed surface roughening by leaching treatment. However, the leaching treatment was not

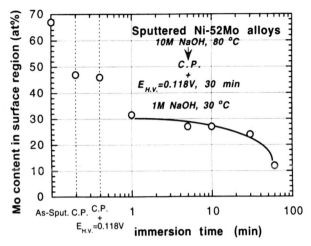

Fig. 4.39. Mo content of surface region for sputter-deposited Ni-52 Mo alloy as a function of immersion time in 10 M NaOH at 80°C. The results of specimens: as-sputtered, pre-polarized at 1 A/m² for 10 min (C.P.) and C.P.+ polarized at the hydrogen overpotential of 0.118 V for 30 min in 1 M NaOH at 30°C, are also shown for comparison

so effective for the alloys with lower molybdenum contents. In the case of the Ni-15Mo alloy, which shows the highest activity, the electrode performance was almost independent of the leaching treatment. This, however, does not mean that the surface activation by the leaching treatment is not effective for the alloys with lower molybdenum contents. The cathodic current density was found to increase with increasing the cathodic potential runs and liable to saturate at the second or third run in 1 M NaOH at 30°C, and hence the cathodic polarization curves shown in Fig. 4.35 are the results at the second or third cathodic polarization. The enhanced activity by repetition of cathodic polarization is significant only for the alloys with low molybdenum contents. Such an improvement in the catalytic activity permits the assumption that in situ leaching of molybdenum occurs during cathodic polarization even in 1 M NaOH at 30°C. This is verified by the reduction of molybdenum content in the surface region measured by XPS, by about 5 at.%, for the Ni-15Mo alloys after the second cathodic polarization. It can, therefore, be said that the leaching treatment enhances the activity by decreasing molybdenum content and increasing surface area.

Ni-Mo alloy electrodes were prepared by mechanical alloying and sputter-deposition methods. The addition of molybdenum to nickel for both electrodes enhances the electrocatalytic activity for HER in 1 M NaOH at 30°C. The alloys prepared by MA show poor electrode performance probably due to PTFE added to the powders as binder to obtain the bulky electrode by cold die-pressing. On the other hand, sputter-deposited Ni-Mo alloy electrodes are found to be active electrocatalysts for HER and the optimum surface

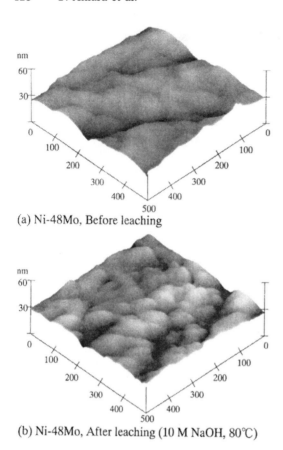

(a) Ni-48Mo, Before leaching

(b) Ni-48Mo, After leaching (10 M NaOH, 80°C)

Fig. 4.40. AFM images of Ni-48Mo alloy before (a) and after (b) leaching treatment in 10 M NaOH at 80°C

composition is about 10 at.% Mo. Its hydrogen exchange current density is more than two orders of magnitude higher than that of nickel metal. Leaching treatment by immersion in 10 M NaOH solution at 80°C significantly improves the activity for Ni-Mo alloys by decreasing molybdenum content to the optimum surface composition and increasing the effective surface area.

4.6 Concluding Remarks

In the present chapter we discussed novel materials obtained by mechanical milling, electrodeposition and sputter deposition methods and those properties. The conclusions are as follows: The cyclic crystalline-to-amorphous phase transformations occurs during ball milling of elemental $Co_{75}Ti_{25}$ powders. An amorphous phase of $Co_{75}Ti_{25}$ is formed in a very short milling time. This amorphous phase does not withstand against the impact and shear forces that are generated by the milling media (balls) and transforms completely into a new metastable bcc-Co_3Ti phase upon milling for 86 ks. It returns to the same amorphous phase of $Co_{75}Ti_{25}$ upon milling for 360 ks. This phase transformation is attributed to the accumulation of several lattice imperfections such as point and lattice defects which raise the free energy from the stable phase to a less stable phase (amorphous).

The tungsten content of the Ni-W electrodeposits is strongly influenced by ammonium chloride concentration in the plating bath and is also influenced by the plating bath temperature. By annealing the Ni- 25.0 at.% W alloy, a significant increase of the hardness occurs, and the Hall-Petch strengthening was observed for the hardness extending to a finest grain size of about 10 nm. When the grain size is less than about 10 nm, the hardness decreases with decreasing the grain size. In the case of the 723 K – annealed specimen having grain sizes about between 5 and 8 nm, the distorted lattice image with about 1 to 2 nm thickness in grain boundary regions was observed by HR-TEM. The decrease in hardness may be due to the significant increase of the intercrystalline volume fraction, especially the fraction associated with the triple junction. The amorphous Ti based alloys were formed by mixing certain amounts of alloying elements. The amorphous Ti based alloys containing B, Si, C or Al alloys yield the superior mechanical properties such as microhardness. It is suggested that the amorphous alloys possess the similar local atomic structure as the corresponding crystalline compounds of TiB, Ti_5Si_3 or TiAl.

Ni-Mo alloy electrodes were prepared both by mechanical alloying and sputter-deposition methods. The alloys prepared by mechanical alloying show poor electrode performance probably due to PTFE added to the powders as binder to obtain the bulky electrode by cold die-pressing. On the other hand, sputter-deposited Ni-Mo alloy electrodes are found to be active electrocatalysts for HER and the optimum surface composition is about 10 at.% Mo. Its hydrogen exchange current density is more than two orders of magnitude higher than that of nickel metal.

Acknowledgments

M. Sherif El-Eskandarany K. Aoki, K. Sumiyama and K. Suzuki would like to thank Mr. S. Ito for his kind supporting during TEM/EDS analyses. We are also indebted to Dr. K. Takada for doing the chemical analyses. Tohru Yamasaki and Yoshikiyo Ogino are deeply grateful to Dr. P. Schloßmacher and Prof. Dr. K. Ehrlich in Forschungszentrum Karlsruhe for useful discussions, and also wish to acknowledge Miss R. Tomohira in Graduate School of Himeji Institute of Technology for her experimental assistance. Asahi Kawashima, Toshio Aihara, Eiji Akiyama, Hiroki Habazaki and Koji Hashimoto would like to thank Mr. Yoshiki Ito for his kind help preparing mechanically alloyed powders, and Prof. Kiyoshi Aoki of Kitami Institute of Technology and Dr. Takejiro Kaneko of IMR Tohoku University for useful discussion. This work has been supported in part by a Grant-in Aid for Scientific Research (A-2 07405032, C-2 09650778) from The Ministry of Education, Science, Sports and Culture, Japan and by a Grant-in-Aid for General Scientific Research (No 06452323) and by the Grand-Aid of ECOMPAS from the Ministry of Japan Education, Science and Cuture.

References

1. C. C. Koch, O. B. Cavin, C. G. MacKamey and J. O. Scarbourgh: Appl. Phys. Lett. **43**, 1017-1019 (1983).
2. C. Politis and W. L. Johnson: J. Appl. Phys. **60**, 1147-1151 (1986).
3. R. Schwarz and C. C. Koch: Appl. Phys. Lett. **49**, 146-148 (1986).
4. E. Hellstern and L. Schultz: J. Appl. Phys. **63**, 1408-1413 (1988).
5. M. Sherif El-Eskandarany, K. Aoki and K. Suzuki: J. Appl. Phys. 71, 2924-2930 (1992).
6. M. Sherif El-Eskandarany, K. Aoki and K. Suzuki: J. Appl. Phys. **72**, 2665-2672 (1992).
7. M. Sherif El-Eskandarany, K. Aoki and K. Suzuki: Appl. Phys. Lett. **60**, 1562-1563 (1992).
8. M. Sherif El-Eskandarany, K. Aoki and K. Suzuki: J. Less Com. Metals **167**, 113-118 (1990).
9. G. Cocco, I. Soletta, L. Battezzati, M. Baricco and S. Enzo: Phil. Mag. **B61**, 473- 486 (1990).
10. M. A. Morris and D. G. Morris: Mater. Sci. Forum **88-90**, 529-536 (1992).
11. B. Huang, R. J. Perez, P. J. Crawford, A. A. Sharif, S. R. Nutt and E. J. Lavernia: Nano Struc. Mat. **5**, 545- 553 (1995).
12. G.-H. Chen, C. Suryanarayana and F. H. Froes: Metallur. Trans. **26A**, 1379-1387 (1995).
13. M. Sherif El-Eskandarany, K. Aoki, K. Sumiyama and K. Suzuki: Appl. Phys. Lett. **70**, 1679-1681 (1997).
14. M. Sherif El-Eskandarany, K. Aoki, K. Sumiyama and K. Suzuki: Scrip. Meta. **36**, 1001-1009 (1997).
15. A. E. Ermakov, E. E. Yurchikov and V. A. Barinov: Phys. Met. Metallov. **52**, 50-58 (1981).
16. ASTM card No. 15-806.
17. A. M. El-Sherik and U. Erb: Plating & Surface Finishing, **82**, 85 (1995).
18. G. Rauscher, V. Rogoll, M. E. Baumgaertner and Ch. J. Raub: Trans. Inst. Metal Finish., **71**, 95 (1993).
19. L. Domnikov: Metal Finishing, **62**, 68 (1964).
20. L. E. Vaaler and M. L. Holt: J. Electrochemical Soc., **90**, 43 (1946).
21. T. Yamasaki, W. Schneider, P. Schloßmacher and K. Ehrlich: Proc. of 2nd Int. Conf. of Micro Materials '97, Berlin, (1997), 654.
22. T. Yamasaki, P. Schloßmacher, K. Ehrlich and Y. Ogino: Proc. of IISMANAM-97, Barcelona 1997, in press.
23. M. Kakudo and N. Kasai: "X-ray Diffraction by Polymers", Elsevier Sci. Pub., 1972.
24. G. D. Hughes, S. D. Smith, C. S. Pande, H. R. Johnson and R. W. Armstrong: Scripta Metall., **20**, 93 (1986).

25. A. Lasalmonie and J. L. Strudel: J. Mater. Sci., **21**, 1837 (1986).
26. W.Y. Lim, E. Sukedai, M. Hida and K. Kaneko: Material Science Forum, **88-90**, 105 (1992).
27. H. Van Swygenhoven, P. Boni, F. Paschoud, M. Victoria and M. Knauss: Nanostructured Materials, **6**, 739 (1995).
28. K. Suzuki, F. Itoh, T. Fukunaga and T. Honda: Proc. 3rd Int. Conf. on Rapidly Quenched Metals, Ed. by B. Canter, The Metals Sociaty, **2**, 410 (1978).
29. G. A. Chadwick and D. A. Smith: "Grain Boundary Structure and Properties", Academic Press, London, 1976.
30. I. A. Ovid'ko: Nanostructured Materials, **8**, 149 (1997).
31. G. Palumbo, S. J. Thorpe and K. T. Aust: Scripta Metall., **24**, 1347 (1990).
32. G. Palumbo, U. Erb and K. T. Aust: Scripta Metall., **24**, 2347 (1990).
33. H. Fujita: Ultramicroscopy, **39**, 369 (1990).
34. H.Van Swygenhoven and A. Caro: Abs. of ISMANAM-97, Barcelona, 1997, 3/4-O-2.
35. M. Naka, K. Hashimoto and T. Masumoto: J. Japan Inst. Met., **38**, 385 (1976).
36. M. Naka, M. Miyake, M. Maeda and I. Okamoto: Scripta Met, **17**, 1293 (1983).
37. M. Naka, M. Miyake, M. Maeda and I. Okamoto: Proc. 5th Int. Conf. on Rapidly Quenched Metals, 1985, 1473.
38. M. Naka, M. Miyake and I. Okamoto: J. Japan Inst. Metals, **5**, 521 (1986).
39. M. Naka, K. Asami, K. Kashimoto and T. Masumoto: Proc. 4th Int. Conf. on Titanium, AIME, 1981, 2695.
40. M. Naka, H. Fujimori and I. Okamoto: Proc. 7th Inter. Conf. on Vacuum Metallurgy, The Iron & Steel Inst. Japan, 1982, 650.
41. K. Hashimoto, Mater. Sci. Eng., **A179/A180**, 27-30 (1994).
42. S. Trasatti, Advances in electrochemical science and engineering, Vol. 2, edited by H. Gerischer and C. W. Tobias, VCH, Weinheim 1992, pp. 1-85.
43. J. Divisek, H. Schmitz and J. Balej, J. Appl. Electrochem., **19**, 519 (1989).
44. I. J. Rat and K. I. Vasu, J. Appl. Electrochem., **20**, 32 (1990).
45. J. -Y. Huot, J. Electrochem. Soc., **136**, 32 (1989).
46. I. Arul Raj and V. K. Venkatesan, Int. J. Hydrogen Energy, **13**, 215-223 (1982).
47. D. E. Brown, M. N. Mahmood, A. K. Turner. S. M. Hall and P. O. Fogarty, Int. J. Hydrogen Energy, **7**, 405 (1982).
48. J.-Y. Huot, M. L. Trudeau and R. Schultz, J. Electrochem. Soc., **138**, 1316 (1991).
49. C. N. Welch and J. O. Snodgross, U. S. Pat. 4251478.
50. D. Miousse, A. Lasia and V. Borck, J. Appl. Electrochem., **25**, 592 (1995).
51. D. E. Brown, M. N. Mahmood, M. C. M. Man and A. K. Turner, Electrochim. Acta, **29**, 1551-1556 (1984).
52. M. Hara, K. Hashimoto and T. Masumoto, J. Appl. Electrochem., **13**, 295 (1983).
53. N. Kumagai, A. Kawashima, K. Asami and K. Hashimoto, Appl. Electrochem., **16**, 565-574 (1986).
54. N. Kumagai, Y. Samata, A. Kawashima, K. Asami and K. Hashimoto, Corrosion, Electrochemistry and Catalysis of Metallic Glasses, R. B. Diegle and K. Hashimoto eds., The Electrochemical Society Pennington N. J., (1988), pp. 390-400.
55. Y. Hayakawa, A. Kawashima, K. Asami and K. Hashimoto, J. Appl. Electrochem., **21**, 1017-1024 (1992).

56. A. Kawashima, T. Kanda and K. Hashimoto, Mater. Sci. Engng., **99**, 521-526 (1988).
57. C. C. Koch, O. B. Cabin, C. G. Mckamey and J. C. Scarbrough, Appl. Phys. Lett., **43**, 1017 (1987).
58. R. B. Schwarz, R. R. Petrich and C. K. Saw, J. Non-Cryst. Solids, **76**, 281 (1985).
59. H. Yoshioka, A. Kawashima, K. Asami and K. Hashimoto, Corrosion, Electrochemistry and Catalysis of Metallic Glasses, R. B. Diegle and K. Hashimoto eds., The Electrochemical Society, Pennington N. J., (1988), pp. 242-253.
60. K. Shimamura, K. Miura, A. Kawashima and K. Hashimoto, Corrosion, Electrochemistry and Catalysis of Metallic Glasses, R. B. Diegle and K. Hashimoto eds., The Electrochemical Society, Pennington N. J., (1988), pp. 232-241.
61. K. Asami, J. Electron Spectrosc., **9**, 469 (1976).
62. K. Asami and K. Hashimoto, Corros. Sci., **17**, 559 (1977).
63. K. Asami, K. Hashimoto and S. Shimodaira, Corros. Sci., **17**, 713 (1977).
64. K. Asami and K. Hashimoto, Corros. Sci., **24**, 83 (1984).
65. K. S. Kim and N. Winograd, Surface Sci, **43**, 625 (1974).
66. B. E. Conway, H. Angerstein-Kozlowska, M. A. Satter and B. V. Tilak, J.Electrochem.Soc., **130**, 1825 (1983).
67. H. Kita, J. Electrochem. Soc., **113**, 1095 (1966).
68. M. Miles and M. Thomason, J. Electrochem. Soc., **123**, 1456 (1976).
69. V. Marian, Ann. Phys. Ser. **11**, 459 (1937).

5 Amorphous and Partially Crystalline Alloys Produced by Rapid Solidification of The Melt in Multicomponent (Si,Ge)-Al-Transition Metals Systems

D. V. Louzguine and A. Inoue

Institute for Materials Research, Tohoku University, Sendai 980-8577, Japan

Summary. The present chapter describes Si,Ge-Al-TM (TM-transition metals) amorphous alloys produced by rapid solidification. Although Al-Si-Fe and Al-Ge-Cr alloys have similar composition ranges for the formation of an amorphous phase in the Al-rich area and the largest possible among the Al-Si,Ge-TM systems metalloid concentration in the amorphous phase achieved, influence of transition metals on the amorphous phase formation between Ge- and Si-based (Si,Ge-Al-TM) alloys is significantly different. Composition range of an amorphous single phase in the Si-Al-Fe system was extended up to 50-60 at % Si by the alloying with Ni, Cr and Zr transition metals while the addition of different transition metals to Ge-Al-Cr alloys with 55-60 % Ge causes appearance of crystallinity. Moreover, microstructure of the Si-based alloys produced by rapid solidification changes from homogeneous amorphous to heterogeneous - (amorphous+crystalline Ge solid solution) by the addition of Ge. The Ge particle size increases with increasing Ge content from 5-7 nm at 7 at % Ge to 30-40 nm at 40 at % Ge. The amount of Si dissolved in Ge decreases with increasing Ge concentration. At the same time replacement of Ge by Si for Ge-Al-Cr-Si causes the precipitation of Ge particles from the amorphous matrix. In contrast to Si-Al-TM-Ge alloys where homogeneous distribution of c-Ge particles in an amorphous matrix was observed, the distribution of the Ge particles in the Ge-Al-Cr-Si alloy is inhomogeneous. This phenomenon as well as the properties of the obtained materials are also discussed in the present chapter.

5.1 Introduction

Fully amorphous alloys containing a relatively large amount of Si or Ge were obtained in the Al-Si-TM and Al-Ge-TM (here TM-3-d transition metals) [1–7] systems by rapid solidification using a melt spinning technique [8] with rapid solidification of the melt on a single copper roller. According to the most comprehensive studies [2,3] these alloys have been found to have wide compositional ranges for the formation of the amorphous single phase. In particular, it has been found that a homogeneously amorphous phase is formed over wide composition ranges of ternary Al-Si-TM (Cr, Mn, Fe, Co and Ni) and Al-Ge-TM (V, Cr, Mn, Fe, Co and Ni) alloys. Moreover, such compositional ranges in the Al-Ge-TM ternary alloys were found to be relatively close to those observed in the Al-Si-TM alloys [2,3]. The widest compositional ranges in the direction of Si or Ge concentration axis were up to 40 at % Si for Al-Si-10 at % Fe and up to 50 at % Ge for Al-Ge-10 at % Cr alloys [3]. Although the compositional ranges of the amorphous phase are almost the same for Al-Si-TM and Al-Ge-TM systems, one can notice a difference, in that the metalloid concentration ranges are somewhat wider for Al-Ge-TM than for Al-Si-TM alloys. As no amorphous phase is formed in binary Al-Si and Al-Ge alloys using the melt spinning technique, it should be noticed that glass-forming ability of Al-Si and Al-Ge alloys is enhanced by the addition of TM.

The mechanical and electrical properties of the Al-Si-TM and Al-Ge-TM alloys change greatly with composition. For example, the room temperature resistivity and Vickers hardness of the Al-Si-TM and Al-Ge-TM alloys increase in the range from 2.2 to 19.4 µΩm and from 200 to 1000 H_V respectively, with increasing Si, Ge or TM content [3].

In the subsequent studies it has been found that an additional alloying with Ni, Cr or Co causes an extension of the compositional range of an amorphous phase in the ternary Al-Si-Fe system up to 45 at % Si. As a result, quaternary amorphous alloys with elevated Si content: $Si_{45}Al_{41}Fe_{10}Cr_4$, $Si_{45}Al_{36}Fe_{15}Cr_4$, $Si_{45}Al_{31}Fe_{20}Ni_4$, $Si_{45}Al_{36}Fe_{15}Ni_4$ and $Si_{45}Al_{41}Fe_{10}Co_4$ were produced [9,10]. These studies established the general possibility of producing the Si-based alloys by the additional alloying.

Amorphous Si, Ge and some of the above alloys are of considerable technological importance in a wide variety of microelectronic, optical and optoelectronic applications. Applications of amorphous semiconductors include xerography, laser printers, electrical switching and memory devices, optical mass memories, photographic imaging, high energy particle detectors, etc. [11]. Si-Ge alloys are used in solar cells and high conversion efficiencies of the cells have been achieved by profiling the band gap of amorphous Si-Ge alloys [12]. On the basis of the above-mentioned it can be said that the production and investigation of new Si- and Ge-based amorphous alloys have both scientific and commercial interest. Amorphous Si, Ge and those alloys, for example SiGe, SiB, SiP, SiSn, SiAl, SiC and some others, have been produced in different laboratories by various methods [11] i.e., thin film deposition from the vapour phase, electrolytic deposition, ion implantation, glow discharge decomposition of SiH_4 and GeH_4, and so on. However, by these methods thin films or wafers are usually produced. The present paper summarizes general features and some properties multicomponent Si- and Ge-based alloys produced by rapid solidification of the melt.

5.2 Multicomponent Fully Amorphous Si and Ge-based Alloys

5.2.1 Influence of Composition and Cooling Rate on the Structure of (Si,Ge)-Al-TM Alloys

As has been noted [9,10], an additional alloying with Ni, Cr or Co causes the extension of the composition range of an amorphous phase in the ternary Al-Si-Fe system up to 45 at % Si. The positive influence of multicomponent alloying on the formation of the amorphous phase has been reported in a number of works [13], for example. It has also been shown that one added component lowers the probability of the formation of any concentration fluctuations in liquid alloys [14]. Mutual alloying with different transition metals leads to better fulfilment of the three empirical rules for achieving high glass forming ability, i.e., [1] multicomponent alloy system, [2] large atomic size ratios among the main constituent elements, and [3] negative heats of mixing among the constituent elements. The Si-Al-TM system satisfies these requirements. It is known that the formation of multicomponent crystalline phases in an equilibrium state enhances the glass-forming ability of the alloy prepared by rapid solidification. Also the presence of several types of atoms in the amorphous phase impedes diffusion and helps to prevent crystallization. The above-mentioned principles can be illustrated on the example of Si-Al-Fe alloys. Simultaneous addition of Ni and Cr results in the expansion of the amorphous single phase area in the direction of the Si concentration axis. At 50 at % Si in the Si-Al-Fe-Ni-Cr system, an amorphous single phase is formed in the composition range of 22-28 at % Al, 10-18 at % Fe, 3-7 at % Ni and 3-7 at % Cr. By the addition of the 4-d transition metal Zr, the concentration range of the amorphous single phase increases up to 55 at % Si (Fig. 5.1).

The typical composition is $Si_{55}Al_{20-22}Fe_{8-10}Ni_{5-7}Cr_{3-5}Zr_{3-5}$. There is no ternary Al-Ge-TM system in which Ge-based amorphous alloys containing at least 50 at % Ge are formed during melt spinning except for the Al-Ge-Cr system. A $Ge_{50}Al_{40}Cr_{10}$ alloy also has an amorphous structure after rapid solidification at the surface velocity of the copper wheel of 42 m/s (see Fig. 5.1). The shape of two broad diffraction peaks is different from that for the Si-based alloys (see Fig. 5.1). In the case of Si-based alloys, two shoulders of the first broad peak closely overlap with each other. Two shoulders of the $Ge_{50}Al_{40}Cr_{10}$ alloy appear to be more separate (see Fig. 5.1) that is also clearly seen in Fig. 5.2. The split of the first halo peak has been interpreted as an overlapping diffraction phenomenon resulting from two separate amorphous phases namely an Al-enriched and a Ge-enriched one [3].

Such a compositional phase separation probably occours as an early stage of spinodal decomposition of the amorphous phase during rapid solidification. For all alloys described the surface velocity of the copper wheel was 42 m/s [15]. Amorphization of the alloy containing 60 % Si was achieved by

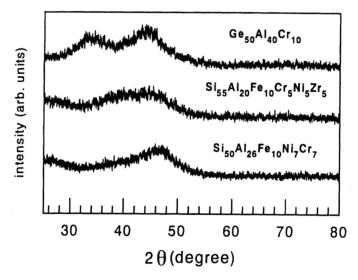

Fig. 5.1. X-ray diffraction patterns of some rapidly solidified Si- and Ge-based alloys

using a surface velocity of the copper wheel of 63 m/s. Alloys with 50 and 55 at % Si were found to have a definite compositional area of homogeneity [15] of the amorphous phase, whereas a slight change in composition of the $Si_{60}Al_{18}Fe_{12}Ni_4Cr_3Zr_3$ alloy led to precipitation of a crystalline phase. Conditions for the formation of the amorphous phase depending on the surface velocity of the copper wheel are summarized in Table 5.1.

The $Si_{50}Al_{26}Fe_{10}Ni_7Cr_7$ alloy exhibits a relatively high glass-forming ability. The amorphous single phase is formed during rapid solidification for circumferential speed of the copper roller from 21 to 63 m/s, which is useful from the commercial point of view.

Table 5.1. Microstructure of Si-based alloys for various cooling rate controlled by the copper wheel surface velocity

Alloy composition	Copper wheel surface velocity, m/s				
	16	21	35.1	42	63
$Si_{50}Al_{26}Fe_{10}Ni_7Cr_7$	am.+ cryst.	am.[1]	am.	am.	am.
$Si_{55}Al_{20}Fe_{10}Ni_5Cr_5Zr_5$	cryst.	am.+ cryst.[2]	am.+ cryst.	am.	am.
$Si_{60}Al_{20}Fe_8Ni_4Cr_4Zr_4$	cryst.	cryst.[3]	cryst.	am.+ cryst.	am.

[1] Fully amorphous
[2] Amorphous + crystalline
[3] Crystalline

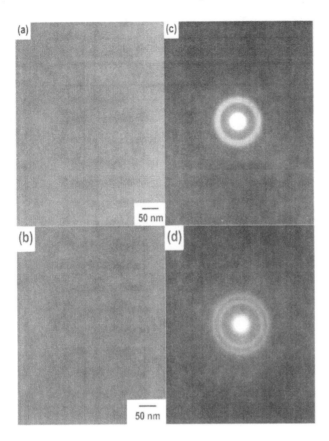

Fig. 5.2. Bright-field electron micrographs (**a** and **b**) and selected-area diffraction patterns (**c** and **d**) of rapidly solidified $Si_{55}Al_{20}Fe_{10}Ni_5Cr_5Zr_5$ and $Ge_{50}Al_{40}Cr_{10}$ alloys

Although Ge belongs to the same group as Si in the periodic table and similar composition ranges for amorphous phase formation were found in the Al-Ge-TM and Al-Si-TM systems, an opposite influence of TM additions has been found in the case of Ge-based alloys. Mutual addition of different transition metals to Ge-Al-Cr system caused the formation of a mixed amorphous +crystalline structure in the rapidly solidified Ge-Al-Cr, Ge-Al-Cr-Mn, Ge-Al-Cr-Fe, Ge-Al-Cr-Fe-Mn, Ge-Al-Cr-Ni-Fe and Ge-Al-Cr-Ni-Fe-Zr alloys [16] containing 55-60 at.% of Ge and various concentrations of transition metals (Table 5.2). No amorphous single phase was obtained for Ge content higher than 50 at %.

In the alloys with 60 at % Ge only a small volume fraction of amorphous phase was observed. As an example, the X-ray diffraction patterns of some alloys having the highest volume fraction of the amorphous phase among the Ge-based alloys, are shown in Fig. 5.3. The volume fraction of

Table 5.2. Compositions of several Ge-based alloys containing 50-60 at % Ge

Composition, at %											Structure
Ge	Al	Cr	Mn	Fe	Ni	Zr	Co	Ti	V	Cu	
50	40	10	-	-	-	-	-	-	-	-	amorphous
50	40		10								amorph.+cryst.[1]
50	40			10							amorph.+cryst.[1]
50	40				10						amorph.+cryst.[1]
50	40					10					amorph.+cryst.[1]
50	40						10				amorph.+cryst.[1]
50	26	7	-	10	7	-	-	-	-	-	amorph.+cryst.[1]
55	33	7	5	-	-	-	-	-	-	-	amorph.+cryst.[1]
55	33		5	7							amorph.+cryst.[1]
55	33	7		5							amorph.+cryst.[1]
55	33	7			5						amorph.+cryst.[1]
55	33	7				5					amorph.+cryst.[1]
55	33	7					5				amorph.+cryst.[1]
55	33	7						5			amorph.+cryst.[1]
55	33	7							5		amorph.+cryst.[1]
55	20	5		10	5	5					amorph.+cryst.[1]
55	25	7	5		8						amorph.+cryst.[1]
55	22	7		8	8						amorph.+cryst.[1]
55	23	7	3		7	5					amorph.+cryst.[1]
60	28	7	5								cryst.+amor.[2]

amorph.+cryst.[1]- amorphous + crystalline
cryst.+amor.[2]- crystalline + amorphous

crystalline phases: Ge, $Fe_{1.5}Ge$ or an unknown phase is low compare to the other alloys studied with 55-60 at % Ge. Among the alloys with 55 at % Ge, the $Ge_{55}Al_{33}Cr_7Mn_5$ alloy shows a diffraction pattern which is the best match with the pattern of an amorphous alloy. However, weak crystalline peaks corresponding to the Ge are also seen (see Fig. 5.3). Thus, in the case of amorphous alloys in the Ge-Al-Cr system, no positive influence of multicomponent alloying with different TM was found. The replacement of Mn by Fe for the $Ge_{55}Al_{32}Cr_7Mn_3Fe_3$ alloy increases the volume fraction of Ge particles and the intensity of the Ge peaks (see Fig. 5.3). Although Fe is an important element for the stabilization of the amorphous phase in the Si-based alloys, in the case of Ge-based alloys an opposite influence of Fe addition is found. As no ternary compounds were formed in the Ge-Al-Fe

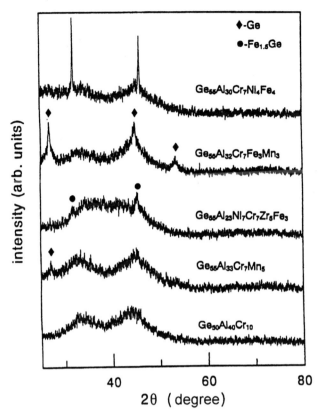

Fig. 5.3. X-ray diffraction patterns of some rapidly solidified Ge-based alloys [16]

system [17], one could instead consider repulsive interaction in this system. The increase in the volume fraction of the Ge particles by the addition of Fe is interpreted to result from the weakening of the bonding among the constituent elements. It follows that the interaction in the Ge-Al-TM system is significantly different from the Si-Al-TM system.

5.2.2 Reasons for the Elevated Glass-forming Ability

In accordance with the rules for achieveing high glass-forming ability, one of the important reasons for the formation of the amorphous phase is related to a large difference in the atomic radii among Si, Al and TM elements (Al, Zr and other TM (Fe, for example) atomic radii are 1.22 1.62 and 1.07 times larger, respectively, than the Si radius). The atomic radius of Al is also about 1.16 times larger than those of Ge and Cr.

From the result [18] that a number of ternary compounds are formed in the Ge-Al-Cr system, it is reasonable to consider that the interaction among Ge, Al and Cr atoms is attractive [19]. Si also produces a number of ternary phases with Al and Fe [20], indicating that the enthalpy of the formation of Al-Si-Fe alloys is large and negative [19], which is favorable for amorphization. Binary systems also show a strong tendency of Al, Si and Fe to produce chemical compounds as is suggested by the large negative heats of mixing among the constituent elements estimated from the data in Table 5.3. Based on the above mentioned concept, it can be explained why the addition

Table 5.3. Enthalpy of formation (DH) (kJ/mol) of binary AlTM and SiTM compounds of equiatomic composition [21]

	Fe	Cr	Ni	Zr
Si	-32	-30	-48	-83
Al	-17	-30	-33	-101

of Al - the element without attractive interaction with Si - is essential for the formation of the Si-rich amorphous phase; if the Al content is lower than 20 at %, no amorphous single phase is formed.

One of another reasons for high glass-forming ability of Si-Al-TM alloys is the complicated structure of the stable $Si_{10}Al_4Fe_2NiCrZr$ compound. This phase has a tetragonal lattice with lattice parameters $a=0.6890$ nm and $c=0.9402$ nm. At the high cooling rate which is realized during melt spinning, the formation of the complicated multicomponent phase is hampered and the liquid becomes an amorphous solid, because of its high viscosity.

5.2.3 Properties

Among the properties of the studied alloys, hardness, crystallization temperature and electrical resistivity have been reported (Table 5.4) [16].

The Si-based amorphous alloys are characterized by elevated electrical resistivity at room temperature in comparison with Al-based amorphous alloys of Al-Si-Fe system [3]. The amorphous matrix in the mixed amorphous+c-Ge

Table 5.4. Some properties of Ge-based amorphous alloys in comparison with Al- and Si-based ones

Alloy	Vickers hardness, H_V	Electrical resistivity, µΩm	Crystallization temperature, K
$Ge_{50}Al_{40}Cr_{10}$	595	8.25	497
$Al_{66-41}Si_{20-45}Fe_{10}TM_4$	550-840	6.2-7.9	550-645
$Si_{50}Al_{26}Fe_{10}Ni_7Cr_7$	926	11.70	723
$Si_{55}Al_{20}Fe_{10}Ni_5Cr_5Zr_5$	935	12.30	701

$Ge_{55}Al_{33}Cr_7Mn_5$ alloy is depleted in Ge that may be responsible for a decrease of hardness (H_V) from 595 to 547 for the $Ge_{50}Al_{40}Cr_{10}$ and $Ge_{55}Al_{33}Cr_7Mn_5$ alloys, respectively. The electrical resistivity also decreases from 8.25 to 7.76 µΩm. Taking in account that the volume fraction of Ge particles is rather low and the interparticular distance exceeds the Ge particle size, we suppose that the decrease in the above-mentioned properties is caused by the change in matrix composition. The $Ge_{50}Al_{40}Cr_{10}$ amorphous alloy is characterized by lower hardness and crystallization temperature in comparison with Al- and Si-based alloys. The temperature dependence of the electrical conductivity of two Si-based alloys shows a slight increase in conductivity and a decrease in resistivity with temperature Fig. 5.4. Such a behaviour is quite common for metallic glasses with relatively high resistivity of more than 1.5 µΩm [22].

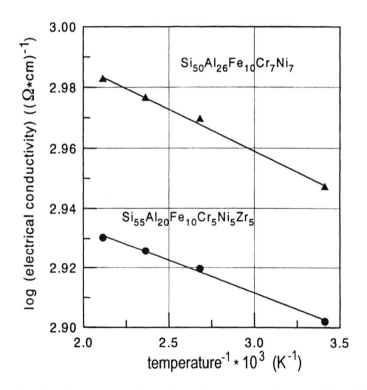

Fig. 5.4. Temperature dependence of the electrical conductivity of Si-Al-TM alloys

5.2.4 Thermal Stability and Crystallization of the Amorphous Phase

Multicomponent Si-based amorphous alloys have a relatively high thermal stability [23]. The crystallization temperatures of the $Si_{50}Al_{26}Fe_{10}Ni_7Cr_7$ and $Si_{55}Al_{20}Fe_{10}Ni_5Cr_5Zr_5$ alloys measured by DSC (Fig. 5.5) at a heating rate of 0.042 K/s were 699 K and 697 K, respectively, which are higher by about 100 K than those for Si_{45}Al-Fe-Cr and Si_{45}Al-Fe-Ni quaternary alloys. The shape of these two curves is greatly different. In the case of the $Si_{55}Al_{20}Fe_{10}Ni_5Cr_5Zr_5$ alloy, only one sharp exothermic peak with high enthalpy of transformation of 295 J/g is observed. This is due to the fact that this alloy is presumed to have a higher degree of dense random packed structure. The same crystallization curve is shown for $Ge_{50}Al_{40}Cr_{10}$ alloy as well. As has been shown in Table 5.4, Ge-based alloys have the lowest crystalliza-

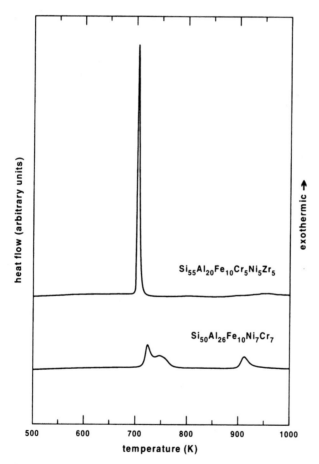

Fig. 5.5. DSC curves of the amorphous $Si_{55}Al_{20}Fe_{10}Ni_5Cr_5Zr_5$ and $Si_{50}Al_{26}Fe_{10}Ni_7Cr_7$ alloys [15]

tion temperature among Si- and Al-based alloys discussed. This is related to the fact that Si and its alloys have a higher melting temperature than Ge and its alloys. Although Al is an element with a relatively low melting temperature, the alloying with TM raises this temperature. According to nanobeam EDX spectroscope analysis of the $Si_{55}Al_{20}Fe_{10}Ni_5Cr_5Zr_5$ alloy in a small volume of the specimen (beam size of 5-25 nm), the chemical composition of crystalline particles ($Si_{10}Al_4Fe_2NiCrZr$) precipitated from the amorphous solid during heat treatment was close to the nominal composition of the alloy. All of the six elements are dissolved into the complicated crystalline phase. As has been stated above, this phase was found to have a tetragonal lattice with lattice parameters $a=0.6890$ nm and $c=0.9402$ nm. No precipitation of the $Si_{10}Al_4Fe_2NiCrZr$ phase was found in the Zr-free $Si_{50}Al_{26}Fe_{10}Ni_7Cr_7$ alloy. The thermal stability of the $Si_{55}Al_{20}Fe_{10}Ni_5Cr_5Zr_5$ alloy was examined by annealing of as-solidified samples at different temperatures. A time-temperature transformation diagram is shown in Fig. 5.6. It is seen that at temperatures lower than 600 K the $Si_{55}Al_{20}Fe_{10}Ni_5Cr_5Zr_5$ alloy is stable for more than 100 hours. No metastable phase was formed at temperatures lower than the crystallization temperature. At higher temperatures a diagram for ribbon samples, as plotted in Fig. 5.6b, was constructed by using the incubation time from isothermal calorimetry curves. The ribbon samples were annealed isothermally at temperatures of 667, 671, 675, and 679 K. Heating up to each annealing temperature was done at a heating rate of 3.3(3) K/s. Points on the curve correspond to less than 0.5% of the volume fraction transformed.

The study of the kinetics of crystallization of the $Si_{55}Al_{20}Fe_{10}Ni_5Cr_5Zr_5$ alloy was performed using the Johnson-Mehl-Avrami isothermal analysis for volume fraction x transformed as a function of time t on the basis of the following equation [24]

$$x(t) = 1 - \exp[-kt^n] \:. \tag{5.1}$$

The samples were annealed isothermally at several temperatures: 667, 671, 675, and 679 K (see Fig. 5.7a). The volume fraction transformed versus the annealing time plot is shown in Fig. 5.7b. The volume fraction transformed at the time t is assumed to scale with the fraction of the total heat released. Equation (1) can be written as

$$\ln\ln[1/(1-x)] = \ln(k) + n\ln(t) \:, \tag{5.2}$$

where k is an effective rate constant and n is the Avrami exponent. The Avrami plot of $\ln[-\ln(1-x)]$ vs. $\ln(t)$ yields a straight line with slope n and intercept $n\ln(k)$. Figure 5.7 c shows this plot at different temperatures. The plots are reasonably linear. There is no systematic variation of slope with temperature. The average slope is $n=3.4$. For polymorphic and eutectoid transformations this value suggests the involvement of an overlap three-dimensional linear growth on quenched-in nuclei and on new nuclei. In differential scanning calorimetry, the temperature at which the maximum deflection is observed

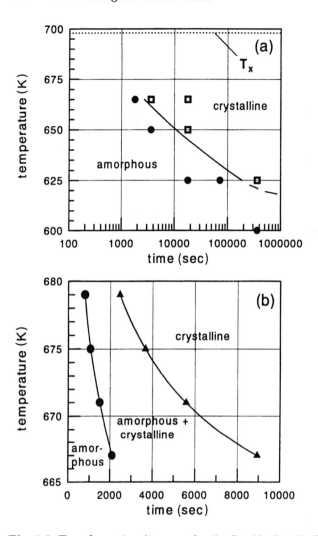

Fig. 5.6. Transformation diagrams for the $Si_{55}Al_{20}Fe_{10}Ni_5Cr_5Zr_5$ alloy at different temperature ranges (**a** and **b**) (T_x- crystallization temperature, measured by DSC at a heating rate of 0.042 K/s) [23]

varies with heating rate. By marking a number of differential thermal patterns at different heating rates, the kinetic constant and an energy barrier opposing the crystallization can be obtained directly from the DSC data. The activation energy E (the energy barrier opposing the crystallization) and a frequency factor A (a measure of the probability that an atom having energy E will participate a reaction) in the equation for the fraction transformed x:

$$dx/dt = A(1-x)\exp(-E/RT) \tag{5.3}$$

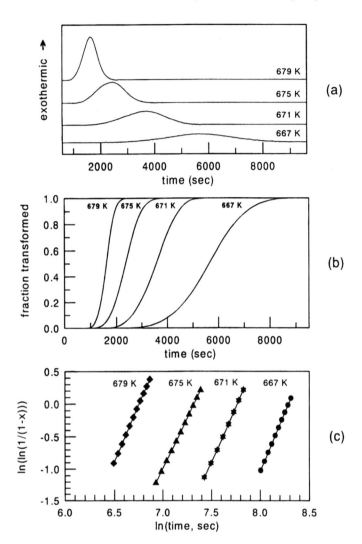

Fig. 5.7. (a) Isothermal differential calorimetry curves, (b) the fraction transformed versus the annealing time, and (c) the Avrami plot for the $Si_{55}Al_{20}Fe_{10}Ni_5Cr_5Zr_5$ alloy [23]

can be obtained directly from the temperature T_m (peak temperature) at which the ratio dx/dt attains the maximum [25]. The corresponding equation is:

$$-E/R = d(\ln V_h/T_m^2)/d(1/T_m) ,\qquad(5.4)$$

where V_h is the heating rate and R is the gas constant.

The data plotted as indicated by this equation i.e., $(\ln V_h/T_m^2)$ vs. $(1/T_m)$, are shown in Figure 5.8a. Heating rates used were 0.042, 0.083, 0.17, 0.67

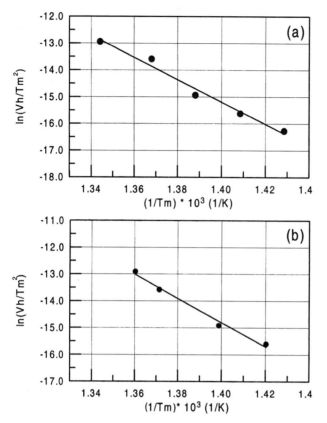

Fig. 5.8. Kissinger plot for (a) $Si_{55}Al_{20}Fe_{10}Ni_5Cr_5Zr_5$ and (b) $Si_{50}Al_{25}Fe_{10}Ni_5Cr_5Zr_5$ alloys [23]

and 1.33 K/s. From the slope, the activation energy was estimated to be 340 kJ/mol (or 3.55 eV). The same Kissinger plot ($\ln V_h/T_m^2$) vs. ($1/T_m$) for the $Si_{50}Al_{25}Fe_{10}Ni_5Cr_5Zr_5$ alloy is shown in Fig. 5.8b. The slope value is close to that for the $Si_{55}Al_{20}Fe_{10}Ni_5Cr_5Zr_5$ alloy and the activation energy is estimated to be 370 kJ/mol.

5.2.5 Production of Bulk Amorphous Samples by Hot Pressing. Densification Behaviour

Si-based amorphous compacts in the shape of cylinders were produced by the consolidation of an amorphous powder [26]. Cylinders of 4 mm in diameter and 2 mm in height ("small" cylinders) were used for the preliminary testing in order to investigate the general possibility of bulk sample production and to evaluate the most suitable heating cycle between isothermal, when pressure is applied after heating to desired temperature and heating-up one, when pressure is applied gradually during heating. Cylinders of 10 mm in diameter and 4 mm in height were used in further studies to determine the critical temperature of pressing i.e. the highest temperature of pressing at which the bulk sample is still amorphous. The consolidation process was performed by using a uniaxial hot-pressing equipment under a vacuum of 10^{-4} Pa. A tungsten carbide hot pressing die was used [27]. About 100 mg of the powder was tapped into a die with an inner diameter of 4 mm and then consolidated in a 2 kN hot-pressing installation. The compacts with a diameter of 10 mm were fabricated using 30 kN hot-pressing equipment. The die was heated at a rate of 0.33 K/s up to the desired temperature, held for 180 s and then cooled by flowing argon gas. Such a method for the amorphous bulk sample production was used for Al- and Fe-based alloys and the best hot-pressing cycle conditions were determined [27]. The pressure was applied gradually and the highest value reaches 1.5 GPa. The amorphous structure of the $Si_{55}Al_{20}Fe_{10}Ni_5Cr_5Zr_5$ remains unchanged after ball milling. On the basis of preliminary production of "small" cylinders, it has been noticed that the heating-up pressing cycle in which pressure is applied during heating is a better method as compared with the isothermal pressing for the consolidation of amorphous powder. In order to produce an amorphous compact at the temperature below the crystallization temperature, the heating-up pressing cycle should be used. The starting temperature for the applying pressure was found to be 150 K less than maximum one. The critical temperature was estimated to be about 697 K.

The temperature dependence of the relative density was also studied for the $Si_{55}Al_{20}Fe_{10}Ni_5Cr_5Zr_5$ alloy produced in the shape of cylinders of 10 mm in diameter and 4 mm in height by hot pressing of the amorphous powder during heating up to various temperatures under a critical pressure of 1.5 GPa using the above-mentioned heating-up pressing cycle. The critical pressure corresponds to the maximum pressure which can be generated by the present technique. The density of compacts increases gradually with increasing temperature and reaches a maximum of 98.3% near the critical temperature of 697 K (Fig. 5.9) which is estimated as an average between 687 and 708 K. This temperature is close to the crystallization temperature of the amorphous ribbon (698 K) measured by DSC at a heating rate of 0.042 K/s. Such a value of relative density is close to the highest densities that can be achieved by using the consolidation of the amorphous powder [27]. The Vickers hardness of the compact 940±25 is nearly the same as that reported for the ribbon samples. It can be seen that the densification is highly promoted at temperatures close to the crystallization temperature by the decrease in strength and the large viscous flow of the amorphous alloy. Yield stress values at several temperatures were calculated by using the equation [28]:

$$\sigma_y = 3P/2\ln[1/(1-D)], \tag{5.5}$$

where σ_y is the yield stress, P is the external pressure and D is the relative density of the compact. The yield stress-temperature diagram is plotted in Fig. 5.10. It is seen that near the crystallization temperature of the amor-

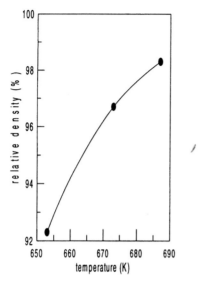

Fig. 5.9. Pressing temperature dependence of the relative density of the $Si_{55}Al_{20}Fe_{10}Ni_5Cr_5Zr_5$ compacts [26]

phous phase a small change in temperature tends to produce a large change in yield stress. The pressing time up to 180-200 s also promotes the densification of the amorphous powder, but the long holding at the critical temperature does not have any influence on the sample density. The exposure to temperatures above 697 K results in the formation of a mixed amorphous+crystalline

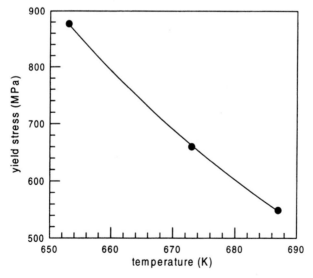

Fig. 5.10. Yield stress-temperature diagram [26]

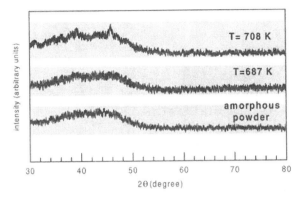

Fig. 5.11. X-ray diffraction patterns of amorphous powder and bulk samples produced at two different temperatures [26]

structure. The X-ray diffraction patterns taken from the amorphous powder and compacts obtained at the temperatures of 687 K and 708 K are shown in Fig. 5.11. Weak crystalline peaks are seen in the X-ray diffraction pattern of the bulk sample hot-pressed at 708 K. The DSC curve of the bulk sample produced at 687 K corresponds to that of the amorphous ribbon (Fig. 5.12).

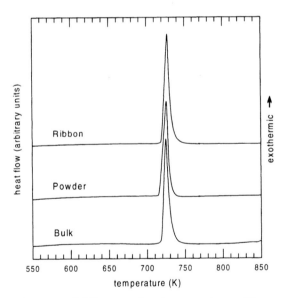

Fig. 5.12. DSC curves for the amorphous ribbon, powder and bulk sample taken at 0.67 K/s [26]

The enthalpy of transformation was found to be 284 J/g, which is very close to that (295 J/g) for the amorphous ribbon. The DSC curve of the powder sample is also given for comparison.

5.3 Precipitation of Nanocrystalline c-Ge Particles in Mixed Si-Ge-Al-TM and Ge-Si-Al-TM Alloys

5.3.1 Microstructure and Phase Composition of Rapidly Solidified Si-Ge-Al-TM Alloys

In alloys, fine scale microstructures are often of importance, and they have become of even greater interest now that nanometer dimensions can be reached. Nanostructured materials have attracted considerable scientific interest recently due to their unusual physical and chemical properties. The finest grain sizes can be achieved when there is copious homogeneous nucleation in the melt. This can be achieved in alloy compositions and processing conditions close to glass formation. Also the crystal nucleation frequency should be high and crystal growth must be slow. Such materials can be produced in Si-Ge-Al-TM alloys.

Although Ge has the same crystalline structure as Si and an unlimited mutual solubility in the solid state is observed between Si and Ge [29], the microstructure of the above-mentioned alloys is sensitive to Ge addition. The microstructure of the Si-Al-TM alloys produced by rapid solidification changes from homogeneous amorphous to heterogeneous – amorphous+c-Ge (c-Ge: crystalline Ge solid solution) by the addition of Ge [30]. The volume fraction of the crystalline phase increases in correspondence to the increase in the Ge content. The microstructure and selected-area electron diffraction pattern of the $Si_{45}Al_{20}Ge_{10}Fe_{10}Ni_5Cr_5Zr_5$ alloy are shown in Fig. 5.13. Sharp rings in the diffraction patterns are produced by the c-Ge particles. The first one corresponds to (111) Ge. The microstructure of the $Si_{48}Al_{20}Fe_{10}Ge_7Ni_5Cr_5Zr_5$ alloy having an amorphous type X-ray diffraction pattern is also inhomogeneous and contains c-Ge particles of less than 5 nm in size embedded in an amorphous matrix (Fig. 5.14) [31]. A broad and very weak peak related to (111) c-Ge is located at $2\theta = 27\text{-}28°$ degrees. Nanoscale crystalline clusters of Ge are also recognized in the structure of $Si_{50}Al_{20}Fe_{10}Ge_5Ni_5Cr_5Zr_5$ alloy, but their density and volume fraction are negligibly low. The c-Ge particle size increases with increasing Ge content and reaches about 36 nm at 40 at % Ge (Fig. 5.15a). The average c-Ge particle size calculated from the broadening of the diffraction peaks at 12-40 at % Ge was found to be in good agreement with that obtained from TEM observations (see Figs. 5.13 and 5.14). The change in the microstructure with Ge content is schematically illustrated in Fig. 5.15b. The homogeneous amorphous structure was formed in the alloys containing up to about 5 at % Ge. The structure consisting of nanocrystalline c-Ge particles homogeneously distributed in an amorphous matrix was observed in alloys containing 7-12 at % Ge. Although

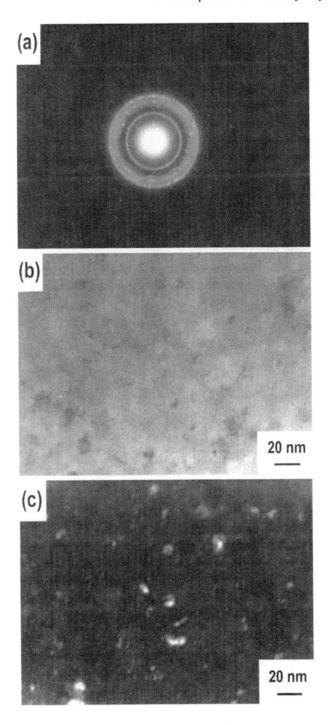

Fig. 5.13. Selected-area electron diffraction pattern; bright and dark-field electron micrographs of rapidly solidified $Si_{45}Al_{20}Ge_{10}Fe_{10}Ni_5Cr_5Zr_5$ alloy (**a–c**), respectively [30]

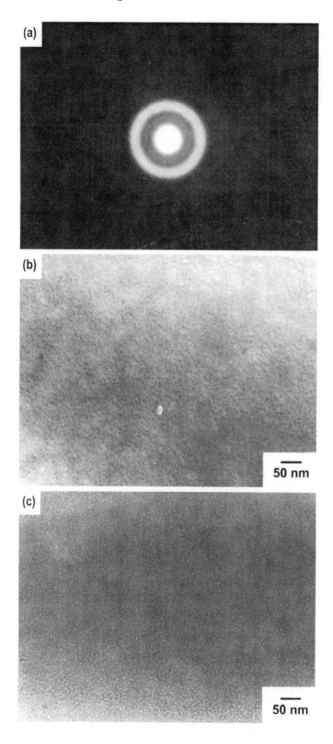

Fig. 5.14. Selected-area electron diffraction pattern (a), bright- (b) and dark-field (c) electron micrographs of rapidly solidified $Si_{48}Al_{20}Fe_{10}Ge_7Ni_5Cr_5Zr_5$ alloy [30]

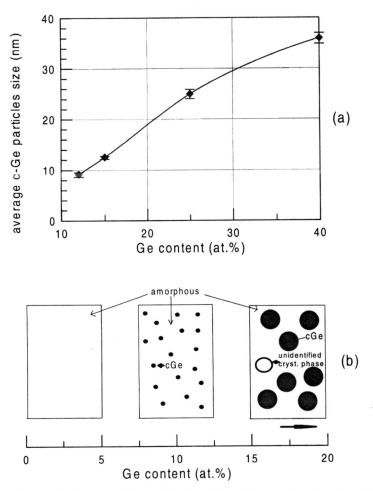

Fig. 5.15. Average c-Ge particle size (a) and changes in microstructure (schematic) (b) as a function of Ge content for $Si_{55-x}Ge_xAl_{20}Fe_{10}Cr_5Ni_5Zr_5$ alloys [30]

nanosize c-Ge particles are also formed in the alloys containing more than 15 at % Ge, another unidentified crystalline phase also appears to crystallize (Fig. 5.16). The X-ray powder diffraction pattern of pure crystalline Ge was used as a reference in order to calculate c-Ge particle size from the broadening of the diffraction peaks. Intrinsic broadening of the diffraction peak β for the cubic lattice can be calculated from the equation [32]:

$$\beta = 0.5B[1 - b/B + (1 - b/B)^{1/2}], \tag{5.6}$$

where b is instrumental broadening and B is the total line broadening of the diffraction peak (rad). The average particle size D_{hkl} in the direction normal to the (hkl) can be calculated from the following equation [2]:

$$D_{hkl} = 0.94\lambda/\beta \cos(\theta), \tag{5.7}$$

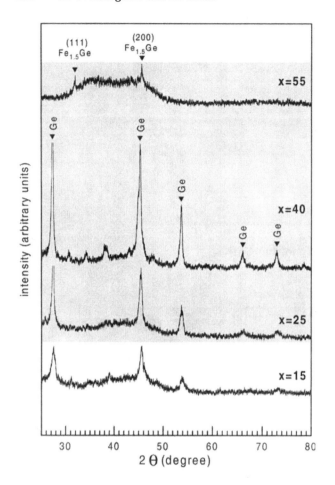

Fig. 5.16. X-ray diffraction patterns of rapidly solidified $Si_{55-x}Ge_xAl_{20}Fe_{10}Ni_5Cr_5Zr_5$ alloys at $x=15$-55 [30]

where λ is the wavelength (nm) and θ is the diffraction angle. The relationships among β_1/β_2, $\cos(\theta_2)/\cos(\theta_1)$ and $\tan(\theta_1)/\tan(\theta_2)$ (1 and 2 are the indexes of peaks) are used to determine the source of the broadening: a small crystal size or microdeformation. If the β_1/β_2 value is close to that of $\cos(\theta_2)/\cos(\theta_1)$, the broadening is caused by nanocrystals. If β_1/β_2 value is between that of $\cos(\theta_2)/\cos(\theta_1)$ and $\tan(\theta_1)/\tan(\theta_2)$, the broadening is caused by crystal size and microdeformation. The part of intrinsic broadening β caused by a small crystal size (n) can be determined by using a plot n/β vs. β_1/β_2 for particles with cubic structure [32]. The results of such a calculation for the $Si_{40}Al_{20}Ge_{15}Fe_{10}Ni_5Cr_5Zr_5$ alloy are shown in Table 5.5 [30]. For the peaks included in Table 5.5, the relationships are: $\beta_{111}/\beta_{220}=0.9378$, $\cos\theta_{220}/\cos\theta_{111}=0.9496$, $\tan\theta_{111}/\tan\theta_{220}=0.5815$. These data indicate that the broadening is mostly caused by the nanosize of the Ge

Table 5.5. Average c-Ge particle size calculated from the broadening of the diffraction peaks

(hkl)	θ(rad.)	b(rad.)	B(rad.)	β(rad.)	D_{hkl}(nm)
111	0.2383	0.003142	0.013963	0.011556	12.9
220	0.3957	0.003840	0.014835	0.011884	13.2
311	0.4689	0.004363	0.017453	0.014102	11.5

particles. Thus, the size and volume fraction of Ge particles in the amorphous matrix are controlled by the Ge concentration in the alloy. The small size and homogeneous distribution of c-Ge particles result from their homogeneous nucleation in the undercooled liquid during rapid solidification. The Ge diffraction peaks are shifted to higher 2θ angles and lower interlattice distances d_{hkl} by the dissolution of Si. The amount of Si dissolved in Ge decreases and the lattice parameter of Ge solid solution increases with increasing Ge concentration in the alloy as shown in Fig. 5.17. The formation of a Ge-Si solid solution argues for the existence of an attractive interaction in the Si-Ge atomic pair. Nanocrystalline Ge solid solution particles are also embedded in an amorphous matrix of the $Si_{51}Al_{18.5}Fe_{9.4}Ge_7Ni_{4.7}Cr_{4.7}Zr_{4.7}$ alloy. In the above-mentioned alloy, the concentration relationship between Si and other constituent elements, except Ge, is equal to that in the Ge-free $Si_{55}Al_{20}Fe_{10}Ni_5Cr_5Zr_5$ alloy and the total metalloid concentration is the highest among Si-based alloys showing an amorphous type X-ray diffraction pattern. The hardness value of the above-mentioned alloy is $H_V=1085$, which is also the highest.

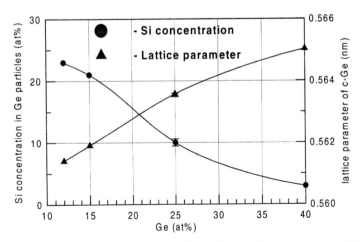

Fig. 5.17. Si concentration in c-Ge and the lattice parameter of Ge solid solution as a function of Ge concentration for $Si_{43-15}Ge_{12-40}Al_{20}Fe_{10}Ni_5Cr_5Zr_5$ alloys [30]

In the X-ray diffraction pattern of the Si-free $Ge_{55}Al_{20}Fe_{10}Ni_5Cr_5Zr_5$ alloy a broad halo peak is also seen (see Fig. 5.16), but the microstructure of this alloy except an amorphous phase contains a small volume fraction of $Fe_{1.5}Ge$ phase with an orthorhombic structure of $a=0.3992$ nm, $b=0.691$ nm, $c=5.00$ nm [33]. It was impossible to obtain a single amorphous phase in the rapidly solidified alloys with 55 at.% Ge at different concentrations of Al, Fe and other transition metals. By the addition of Ge to the $Si_{55}Al_{20}Fe_{10}Ni_5Cr_5Zr_5$ alloy, the electrical resistivity and hardness shown in Fig. 5.18 change slightly in the range up to 15 at.% Ge. The increase in hardness due to the dispersion of nanoscale c-Ge particles is presumed to result from the homogeneous dispersion of the nanoscale c-Ge particles without internal defects which acts as an effective barrier against shear deformation of the amorphous matrix because the particle size is smaller than the size (10 to 20 nm) [30] of the shear deformation band. As has been mentioned, Ge has the same crystal structure as Si and an unlimited mutual solubility in the solid state is observed between Si and Ge. Furthermore, nearly the same compositional ranges in which an amorphous single phase is formed by melt spinning have been observed in the Al-Si-TM and Al-Ge-TM systems [3]. However, the microstructure of the Si-Al-TM-Ge alloys was found to change from homogeneously amorphous to amorphous+c-Ge by the addition of Ge. This result seems to reflect the complicated interaction among Ge, Al, Si and TM atoms in the Si-Al-TM-Ge alloys. As has been mentioned above, Si produces a number of ternary compound phases with Al and Fe, indicating that the heats of mixing among Al,

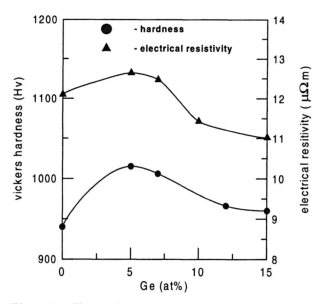

Fig. 5.18. Electrical resistivity and hardness as a function of Ge content for $Si_{55-x}Ge_xAl_{20}Fe_{10}Ni_5Cr_5Zr_5$ alloys [30]

Si and Fe elements are large and negative [19]. In contrast, in the Ge-Al-Fe system no ternary compound phase is formed [17], indicating either low negative heats of mixing for ternary alloy formation or even positive ones. In binary alloys, the enthalpy of formation of Si-TM compounds is always more negative than that of the GeTM compounds as shown in Table 5.6.

Table 5.6. Enthalpy of formation (ΔH, kJ/mol) of GeTM and SiTM compounds [21]

	TMAl, TMSi and TMGe				TMAl$_3$, TMSi$_3$ and TMGe$_3$			
	Fe	Cr	Ni	Zr	Fe	Cr	Ni	Zr
Si	-26	-30	-33	-101	-1	-3	-5	-59
Ge	-9	-13	-20	-97	+6	+4	0	-56

That is one of the reasons why Ge is excluded from the amorphous matrix and precipitates as the Ge-Si solid solution. It follows that in the case of rapid solidification the relatively weak attractive interaction in the Ge-Si atomic pair is obscured by the strong bonding nature among Si, Al and TM atoms. Thus, the exclusion of Ge from the amorphous phase results from the complicated interaction of all atoms involved. Such a material consisting of the nanosize particles of crystalline Ge solid solution (semiconductor with high electrical resistivity of about 5×10^5 μΩm) in the Si-based amorphous matrix with resistivity of 11-13 μΩm depending on its composition can possibly be applied as an electronic material. From the microstructural standpoint, such a new material can be thought of as a composite material in which the size and volume fraction of the crystalline particles embedded in the amorphous matrix are strongly modified by the Ge content in the alloy.

5.3.2 Crystallization Process in the Rapidly Solidified Si-Ge-Al-TM Alloys

The crystallization behaviour of the $Si_{55}Al_{20}Fe_{10}Ni_5Cr_5Zr_5$ alloy also changes upon the addition of Ge. The second exothermic peak due to the precipitation of the Si-Ge solid solution appears (Fig. 5.19) during DSC. The multicomponent $Si_{10}Al_4Fe_2NiCrZr$ phase has the same tetragonal crystalline structure with lattice parameters $a=0.6890$ nm and $c=0.9402$ nm as in the Ge-free $Si_{55}Al_{20}Fe_{10}Ni_5Cr_5Zr_5$ alloy, but has slightly reduced Si content. In a fully crystallized state, the Ge concentration in the Si-Ge solid solution, calculated from the varying of its lattice parameter, changes in the range from 9.5 to 44 at % for the $Si_{50-40}Ge_{5-15}Al_{20}Fe_{10}Ni_5Cr_5Zr_5$ alloys, respectively. The

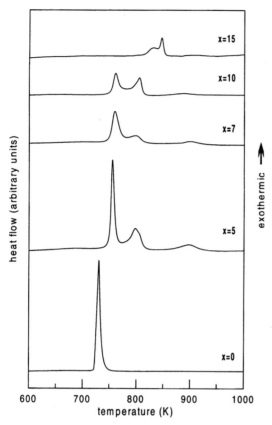

Fig. 5.19. DSC curves of rapidly solidified $Si_{55-x}Ge_xAl_{20}Fe_{10}Ni_5Cr_5Zr_5$ alloys taken at a heating rate of 0.67 K/s [30]

last weak peak on the DSC curves of the Ge-bearing alloys is related to the precipitation of a small amount of the $FeGe_2$ phase having a tetragonal structure with lattice parameters $a=0.5908$ nm and $c=0.4957$ nm [34]. It should be noted that this phase disappears in the Ge concentration range above 10 at %. Thus, the microstructure of the Ge-containing alloys at 5-10 at % Ge in a fully crystallized state consists of a primary crystallized multicomponent SiAlFeNiCrZr phase with a tetragonal structure, cubic Fd3m Si-Ge solid solution and a small amount of I4/mcm $FeGe_2$ phase, as illustrated in Fig. 5.20. The crystallization temperature increases with increasing Ge content and reaches 767 K at 25 at % Ge (Fig. 5.21). The change in matrix composition is the main reason for the rise of crystallization temperature. The addition of Ge causes an increase in the heat of crystallization (total area under DSC peaks see Fig. 5.20) of the rapidly solidified alloys in the range up to 5 at % Ge, followed by the decrease of this value at higher Ge contents in accordance with the decrease in the relative quantity of amorphous phase.

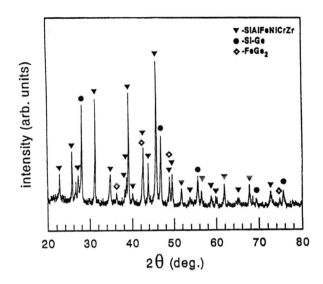

Fig. 5.20. X-ray diffraction pattern of a fully crystallized $Si_{48}Al_{20}Fe_{10}Ge_7Ni_5Cr_5Zr_5$ alloy [30]

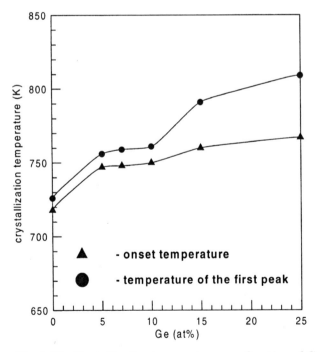

Fig. 5.21. Crystallization temperature as a function of Ge content for $Si_{55-x}Ge_xAl_{20}Fe_{10}Ni_5Cr_5Zr_5$ alloys [30]

5.3.3 The Effect of Si Addition to Melt Spun Ge-Al-TM Alloys

As in the case of Si-Al-TM-Ge alloys, the formation of the Ge-based amorphous alloys was found to be extremely sensitive to Si addition. The replacement of Ge by Si for $Ge_{50-x}Al_{40}Cr_{10}Si_x$ causes the precipitation of Ge particles from the amorphous matrix (Fig. 5.22) even at 3 at % Si and strong diffraction peaks of Ge are seen [16]. The further increase of Si concentration results in an increase of the peak intensity of Ge, the formation of some $Al_{10.7}CrGe_{1.3}$ and another crystalline phase and the disappearance of amorphous phase at more than about 9 at % Si. The structure of the alloys containing less than 2 at % of Ge was found to be homogeneously amorphous. In contrast to Si-Al-TM-Ge alloys, where a fairly homogeneous distribution of c-Ge particles in an amorphous matrix was observed, the distribution of c-Ge particles in the $Ge_{47}Al_{40}Cr_{10}Si_3$ alloy is inhomogeneous. Some structurally different areas were observed. In some parts of the specimen relatively small and homogeneously distributed particles of c-Ge were observed in amorphous matrix. Such an area is shown in Fig. 5.23. The diffraction rings produced by c-Ge are sharp. Figure 5.24 shows c-Ge particles which are much larger in

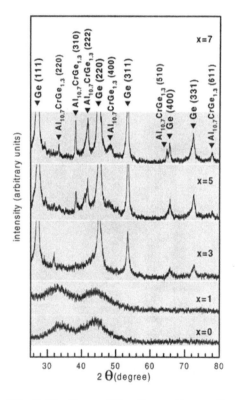

Fig. 5.22. X-ray diffraction patterns of as rapidly solidified $Ge_{50-x}Al_{40}Cr_{10}Si_x$ alloys [16]

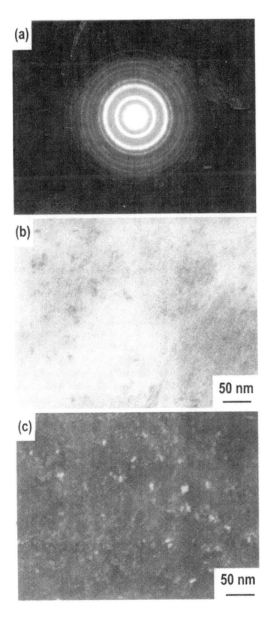

Fig. 5.23. Selected-area electron diffraction pattern (**a**), bright-field (**b**) and dark-field (**c**) electron micrographs of rapidly solidified $Ge_{47}Al_{40}Cr_{10}Si_3$ alloy

size than in Fig. 5.23. The volume fraction of the particles is also higher. As was confirmed by TEM observation, the c-Ge particles have no preferential orientation with respect to the ribbon's surface. Different, even high-indexed zone axes were observed by analyzing the images of the reciprocal lattice (Fig. 5.24c,d). The crystallization temperature and the heat of crystallization were measured by DSC. The corresponding curves are shown in Fig. 5.25. The rest of the amorphous phase in $Ge_{47-43}Al_{40}Cr_{10}Si_{3-7}$ is enriched by Si. After crystallization, Si is incorporated in the $(Si,Al)_2Cr$ phase with a hexagonal structure of $a=0.4496$ nm and $c=0.6377$ nm. It reflects the weakening of bonding nature among Ge, Al and Cr atoms by the addition of Si. It is suggested that the type of atomic interaction in the Ge-Al-Cr system changes from attractive to repulsive by the addition of Si, and that the attractive interaction among Si, Al and Cr atoms is much stronger than that among Ge, Al and Cr atoms.

The compositional ranges for the formation of the amorphous and amorphous+crystalline c-Ge phase in the Si-Al-TM-Ge and Ge-Al-TM-Si systems

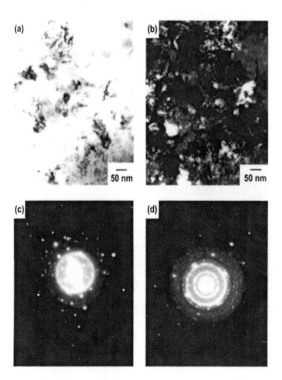

Fig. 5.24. Bright-field (**a**), dark-field (**b**) electron micrographs and selected-area electron diffraction patterns (**c, d**) taken from c-Ge particles of rapidly solidified $Ge_{47}Al_{40}Cr_{10}Si_3$ alloy

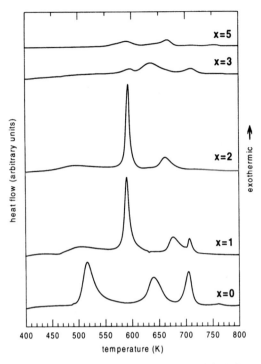

Fig. 5.25. DSC curves of rapidly solidified $Ge_{50-x}Al_{40}Cr_{10}Si_x$ alloys

are summarized in Table 5.7. It can be seen that amorphous and amorphous+nanocrystalline particle structures can be obtained in definite composition ranges of all the systems studied.

Table 5.7. Structures that can be obtained in rapidly solidified Si, Ge, Al and TM containing alloys according to Si and Ge content

System / Structure	Si-Al-TM and Si-Al-TM-Ge	Ge-Al-TM and Ge-Al-TM-Si
Amorphous	50-60%Si[1] (< 5% Ge)	50% Ge (<2% Si)
Amorphous + nanocrystalline c-Ge (5-20 nm)	40-55% Si, 7-15 % Ge	55% Ge
Amorphous + crystalline c-Ge (20-100 nm)	30-5% Si, 25-50% Ge	47-45% Ge, 3-5% Si

[1] atomic percent

References

1. 1. R. O. Suzuki, Y. Komatsu, K. E. Kobayashi, P. H. Shingu: J. Mater. Sci. **18**, 1195 (1983).
2. A. Inoue, M. Yamamoto, H. M. Kimura , T. Masumoto: J. Mater. Sci. Lett. **6**, 194 (1987).
3. A. Inoue, Y. Bizen, H. M. Kimura and T. Masumoto: J. Mater. Sci. **23**, 3640 (1988).
4. K. Dini, R. A. Dunlap: J. Phys. F: Met. Phys. **16**, 1917 (1986).
5. R. A. Dunlap and K. Dini: J. Mater. Res. **1**, 415 (1986).
6. H. S. Chen and C. H. Chen, Phys. Rev. B **33**, 668 (1986).
7. H. M. Kimura, A. Inoue, Y. Bizen, T. Masumoto, H. S. Chen: Mater. Sci. Eng., **99**, 449 (1988).
8. R. Zallen: The physics of amorphous solids (Wiley, New York, 1983), p. 74.
9. D. V. Louzguine, A. Takeuchi, A. Inoue: Mater. Trans., JIM, **38**, 595 (1997).
10. D. V. Louzguine, A. Takeuchi, A. Inoue: J. Mater. Sci. Letters, **17**, 1439-1442 (1988).
11. Electronic and structural properties of amorphous semiconductors, ed by P.G. le Comber and J. Mort Academic press, London, (1973) p. 621.
12. J. R. A. Carlsson, X. H. Li, S. F. Gong, H. T. G. Hentzell: J. Appl. Phys. **74**, 891 (1993).
13. A. Inoue: Mater. Trans., JIM, **36**, 866 (1995).
14. P. J. Desre: Mater. Trans., JIM, **38**, 583 (1997).
15. D. V. Louzguine and A. Inoue: Materials Transactions, JIM, **38**, 1095 (1997).
16. D. V. Louzguine and A. Inoue: Scripta Materialia, **38**, 1280 (1998).
17. Ternary alloys. A comprehensive compendium of evaluated constitutional data and phase diagrams. Edited by G. Petzov and G. Effenberg, V5, (1992). Powder diffraction file, inorganic and organic. International centre for diffraction data V.20-45, (1970-1995) p. 2303.
18. V. V. Milyan and Yu. B. Kuz'ma, Russian Metallurgy (translated from Izvestia Akademii Nauk SSSR, Metally) **4**, 191 (1987).
19. R. de Boer, R. Boom, W. C. M. Mattens, A. R. Miedema, A. K. Niessen: Cohesion in Metals, Elsevier Science Publishers, North-Holland, Amsterdam (1988), p. 37.
20. Ternary alloys. A comprehensive compendium of evaluated constitutional data and phase diagrams. Edited by G. Petzov and G. Effenberg, V5, (1992). Powder diffraction file, inorganic and organic. International centre for diffraction data V.20-45, (1970-1995) p. 2002.
21. R. de Boer, R. Boom, W. C. M. Mattens, A. R. Miedema and A. K. Niessen: Cohesion in Metals, Elsevier Science Publishers, North-Holland, Amsterdam (1988), p. 367.
22. Amorphous Metals ed. H. Matyja, P.G. Zielinski World Scientific Publishing Co, Singapure, (1985) p. 134.
23. D. V. Louzguine and A. Inoue: Materials Research Bulletin, **34**, 1165 (1999).
24. J. W. Christian, The Theory of Transformations in Metals and Alloys, Pergamon Press Ltd., Oxford (1975).
25. H. E. Kissinger, J. Res. National. Bureau Stand. **57**, 217 (1956).
26. D. V. Louzguine, Y. Kawamura, A. Inoue: Materials Science and Technology, **15**, 583 (1999).

27. Y. Kawamura, A. Inoue, A. Kojima, T. Masumoto: in Proc. of the World Cong. on Powder Met. "Adv. in Powder metallurgy" Washington, DC June, Ed. T. M. Cadle and K. S. Narasimhan, Metal Powder Industries Federation, Princeton, 5, 1996, 159-169.
28. A. S. Helle, K. E. Easterling, M. F. Ashby: Acta Metall., **33**, 2163 (1985).
29. Binary Alloy Phase Diagrams, ed by T.B. Massalski, ASM International, Materials Park, Ohio (1990) p. 257.
30. D. V. Louzguine and A. Inoue: Materials Transactions, JIM, **39**, 245 (1998).
31. D. V. Louzguine and A. Inoue: Proc. 4-th Int. Conf. IUMRS-ICA-97 (16-18 Sept. 1997, Mat. Res. Soc. Chiba, Tokyo), the special issue of the Journal NanoStructured Materials, **8**, 1007 (1997).
32. S. S. Gorelik, U. A. Skakov, L. N. Rastorguev: X-ray and Electron-optic Analysis, (in Russian), Moscow, MISIS, (1994), p. 328.
33. A. Adelson and M. Austin: J. Phys. Chem. Solids, **26**, 1795 (1965).
34. E.E. Havinda, H. Damsma, P. Hokkeling: J. Less-Common Met., **27**, 169 (1972).
35. T. Masumoto and R. Maddin, Acta Metall. **19**, 725 (1979).

6 Global CO_2 Recycling – Novel Materials, Reduction of CO_2 Emissions, and Prospects

K. Hashimoto[1], K. Izumiya[2], K. Fujimura[3], M. Yamasaki[3], E. Akiyama[4],
H. Habazaki[3], A. Kawashima[3], K. Asmi[3], K. Shimamura[2], and N. Kumagai[5]

[1]Tohoku Institute of Technology, Sendai, 982-8588, Japan
[2]Mitsui Engineering & Shipbuilding Co., Ltd., Ichihara, Chiba, 290-0067, Japan
[3]Institute for Materials Research, Tohoku University, Sendai, 980-8577, Japan
[4]National Institute of Metals, Tsukuba, 305-0047, Japan
[5]Daiki Engineering Co., Ltd., Kashiwa, Chiba, 277-0804, Japan

Summary. CO_2 emissions, which induce global warming, increase with economic growth. It is impossible to demand that CO_2 emission should be reduced by suppressing economic activity. Global CO_2 recycling can solve this problem. The global CO_2 recycling involves three geographical regions: The electricity is generated by solar cells on deserts. At coasts close to the deserts, the electricity generated on the deserts is used for H_2 production by seawater electrolysis and H_2 is used for CH_4 production by the reaction with CO_2. CH_4 is liquefied and transported to energy consuming districts where, after the CH_4 is used as a fuel, CO_2 is recovered, liquefied and transported to the coasts close to the deserts. Since 90% of city gases in Japan are liquefied natural gas (LNG) consisting mostly of CH_4, CH_4 produced in the global CO_2 recycling can be used immediately as city gas. A CO_2 recycling plant for substantiation of our idea to solve global warming and energy problems has been built on the roof of our institute (IMR) in 1996. Key materials necessary for the global CO_2 recycling are the anode and cathode for seawater electrolysis and the catalyst for CO_2 conversion. All of them have been tailored by us. Since the quantities of CO_2 to be converted far exceed an industrial level, the system must be very simple, the rate of conversion must be very fast and precious metals must not be required for the system. All these requirements are satisfied in the global CO_2 recycling. When the global CO_2 recycling is conducted on a large scale, the energies and costs required to form liquefied CH_4 in the global CO_2 recycling are almost the same as those for production of LNG from currently operating natural gas sources. The project for the field experiment on global CO_2 recycling using pilot plants in Egypt was planned in cooperation with Egyptian scientists, engineers and industries.

6.1 Introduction

The IPCC (Intergovernmental Panel on Climate Change), in August 1990 after the 4th meeting held in Sweden, issued a statement declaring that more than 60% of CO_2 emissions must be cut in order to maintain the level of long-lived greenhouse gases such as CO_2 at the level in 1990 [1]. The CO_2 emissions increase with economic development. Figure 6.1 shows the history and projections of the world total CO_2 emissions by region [2]. The recent

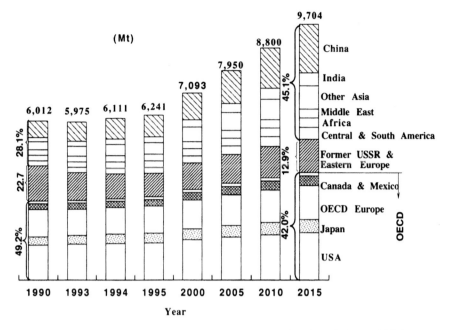

Fig. 6.1. History and projections of CO_2 emissions for the whole world [2]

total CO_2 emissions in the whole world is more than 6 gigatons/year in terms of the weight of carbon. It is reasonable that the share of the developing countries increases. Since it is impossible to decrease economic activity, it is also impossible to decrease CO_2 emissions only by efforts to save and efficiently use of energy. Consequently, CO_2 must be recovered and methods of treatment of recovered CO_2 must be developed.

6.2 Global CO_2 Recycling

The authors are proposing global CO_2 recycling which is able not only to prevent global warming but also to provide abundant renewable energy [3]. As mentioned above about 6 gigatons of CO_2 in terms of weight of carbon is anthropogenically emitted in the whole world in a year. Under the assumptions of a solar constant of 2 cal min^{-1} cm^{-2} and operation of solar cells for 8 hours day^{-1} with a 10% energy efficiency on deserts, it is possible to calculate the area of desert needed to obtain the electricity corresponding to the total energy of combustion of 6 gigatons of carbon in a year. The necessary area is only 0.83% of the sum of the main desert areas on the Earth, that is, only 10.35% of the main desert area of Arabian Peninsula and equals 1.340×10^7 ha. This evidences how rich the Sun is as an energy source. The problem of generated electric power is the difficulty of long-distance transportation exceeding 1000 km and hence other types of energy carriers are required.

One possible use of the electric power generated by solar cells set in desert areas is for the electrolytic generation of H_2 from seawater at a plant built at a nearby coast; the generated H_2 might be liquefied and transported by tankers to energy consumers. However, there are as yet no widely used H_2 combustion systems. The development and distribution of H_2 combustion systems are a great burden on our society in addition to the large energy consumption for liquefaction of H_2 and the low energy of H_2 per volume. It is, therefore, desirable to synthesize some conventional fuel substances from gaseous H_2. Methane, CH_4, is easily synthesized from CO_2 with H_2 through an exothermic reaction:

$$CO_2 + 4H_2 \rightarrow CH_4 + 2H_2O. \tag{6.1}$$

CH_4 is the main component of liquefied natural gas (LNG), and 90% of city gases in Japan are LNG. Accordingly, if liquefied CH_4 is transported by tankers to Japan, CH_4 can be used widely without any change in the present combustion systems. Figure 6.2 shows the scheme of global CO_2 recycling. Solar panels installed in desert areas would generate electricity and this electricity could be used to operate a seawater electrolysis plant at a nearby coast to yield H_2. A CH_4 synthesis plant combined with the electrolysis plant could also be operated to produce CH_4 from the generated H_2 and CO_2 which was recoverd at energy consuming districts of the world and transported by tanker

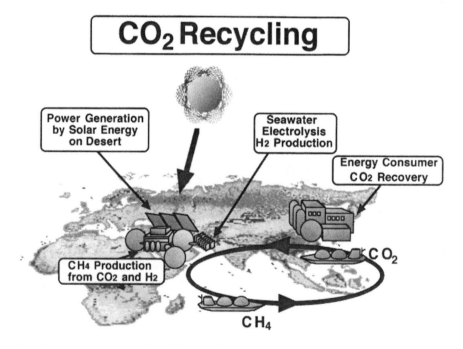

Fig. 6.2. A scheme of global CO_2 recycling

in liquefied form. The produced CH_4 could be liquefied and transported back to the energy consuming districts. Consequently, global CO_2 recycling is an ideal method not only for preventing global warming but also for supplying abundant energy using solar energy, which is expected to last for 5 billion years.

6.3 Key Materials for Global CO_2 Recycling

6.3.1 Cathode Materials

Quite a significant amount of cathode material is needed for the seawater electrolysis plant under consideration to match the huge amount of CO_2 to be processed. Thus, employment of precious metal electrodes for such a plant is unacceptable, although precious metals have the lowest overpotential for hydrogen evolution. Conventional metals having low overpotentials for hydrogen evolution include cobalt and nickel. An attempt was made to decrease the hydrogen overpotential of nickel and cobalt by alloying with various elements and amorphization, using the advantage that amorphous alloys become single phase solid solutions even if various elements are added exceeding their solubility limits at equilibrium.

The hydrogen overpotential of cobalt is slightly lower than that of nickel. The hydrogen overpotential is known to be related to the boiling point of metals, showing a minimum for precious metals such as palladium, rhodium and platinum [4]. This is due to the fact that the boiling point of a metal is strongly related to the bond strength between metal and hydrogen atoms adsorbed on the metal surface, the hydrogen being formed as an intermediate species in the reduction of protons to hydrogen gas. Because the boiling points of cobalt and nickel are lower than those of precious metals, the alloying of cobalt and nickel with refractory elements may decrease the hydrogen overpotential of cobalt and nickel. Accordingly, a refractory element, molybdenum was added to cobalt and nickel.

Co-Mo alloys were prepared by sputter deposition [5]. Their X-ray diffraction patterns are shown in Fig. 6.3. The binary Co-Mo alloys containing 32–63 at % molybdenum show the halo patterns typical of the amorphous structure. The alloys with 27 at % or less molybdenum show diffuse diffraction patterns of the hcp structure, indicating the formation of supersaturated solid solution. Similarly, sputter-deposited Ni-Mo alloys containing 21 at % or more molybdenum became a single amorphous phase and the Ni-Mo alloys with 15 at % or less molybdenum consisted of nanocrystalline fcc solid solution [6]. Consequently, these sputter-deposited binary Co-Mo and Ni-Mo alloys are characterized by the fact that they are composed of a single solid solution phase. The hydrogen evolution reaction on these alloys was examined in 1 M NaOH solution at 30°C by measuring potentiodynamic and galvanostatic cathodic polarization curves. The potential sweep rate for potentiodynamic polarization was 20 mV min^{-1}. Correction for ohmic drop was made by an

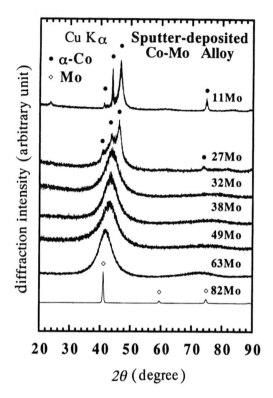

Fig. 6.3. X-ray diffraction patterns of sputter-deposited Co-Mo alloys [5]

IR compensation method for potentiodynamic polarization and by a current interruption method for galvanostatic polarization. Figures 6.4 and 6.5 show cathodic polarization curves of Co-Mo and Ni-Mo alloys. The cathodic polarization curves of cobalt and nickel metals are also included in the figures for comparison.

It can clearly be seen that alloying with molybdenum significantly decreases the overpotential of cobalt and nickel for hydrogen evolution. In particular, a small addition of molybdenum to nickel and cobalt markedly decreases the overpotential, while excess addition of molybdenum rather decreases the beneficial effect of molybdenum addition. For instance, the overpotentials of Ni-15Mo and Co-17Mo alloys for hydrogen evolution at 10^3 A m^{-2} in 1 M NaOH at 30°C are 80 and 175 mV, respectively, whereas those of nickel and cobalt are 400 and 425 mV, respectively.

These polarization curves show two Tafel regions indicating the change in mechanism of hydrogen evolution. The Tafel slopes in the low and high overpotential regions vary in the range 30–50 mV decade^{-1} and 120–140 mV decade^{-1}, respectively. The hydrogen evolution reaction on nickel seems to occur by a series of reactions involving the formation of adsorbed hydrogen

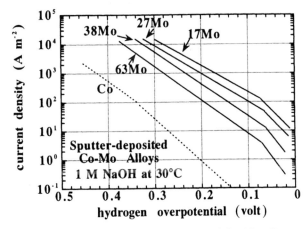

Fig. 6.4. Cathodic polarization curves of Co-Mo alloys measured in a deaerated 1 M NaOH at 30°C [5]

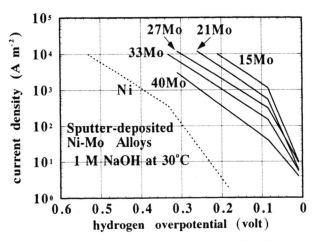

Fig. 6.5. Cathodic polarization curves of Ni-Mo alloys measured in a deaerated 1 M NaOH at 30°C [6]

by discharge of hydrogen ions and subsequent electrochemical desorption of adsorbed hydrogen as shown in the following fomulae [7]

$$H^+ + e \rightarrow H_{ads} \tag{6.2}$$

$$H_{ads} + H^+ + e \rightarrow H_2. \tag{6.3}$$

The rate-determining step seems to be the reaction (6.3), that is, desorption of adsorbed hydrogen. The two Tafel regions are assumed to correspond to the change in the mechanism from low coverage to high coverage of adsorbed hydrogen on the metal surface. As can be seen in Figs. 6.4 and 6.5 the decrease in the hydrogen overpotential by alloy composition results in an increase in the current density in the low Tafel slope region, that is, the current density where the hydrogen coverage changes from almost zero to almost unity. Consequently, the enhancement of the activity for hydrogen evolution facilitates desorption of adsorbed hydrogen.

Figure 6.6 shows the effect of composition of Ni-Mo alloys on the overpotential of hydrogen evolution. The activity of Ni-Mo alloys for hydrogen evolution is remarkably higher than that of alloy component metals. The maximum activity appears at the Ni-15Mo alloy electrode, which is higher than that of the catalytically active platinum electrode. Since the activity of

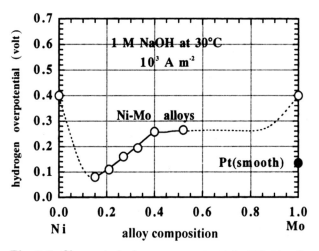

Fig. 6.6. Change in hydrogen overpotential of Ni-Mo alloys at 10^3 A m^{-2} in 1 M NaOH at 30°C with concentration of molybdenum in the alloys [6]

Co-Mo alloys for hydrogen evolution was lower than that of Ni-Mo alloys, an increase in effective surface area by leaching of aluminum from Co-Al and Co-Mo-Al alloys was attempted. Leaching treatment of sputter-deposited alloys was carried out in 1 M NaOH at 30°C until hydrogen bubbling ceased, and the roughness factor was determined electrochemically. The roughness factor of the chemically etched nickel was about 2, and those of cobalt alloys were in the range from 6 to 1100, depending on the alloy composition. The improvement of the activity for hydrogen evolution was partly based on the increase in the effective surface area. However, synergistic effects of alloy constituents is another factor contributing to high catalytic activity. True polarization curves were estimated from the observed current and the real surface area obtained from the roughness factor.

Figure 6.7 shows the variation of hydrogen overpotential at a constant current density estimated from polarization curves and real surface area [5]. It is clear that the additions of molybdenum and aluminum to cobalt enhance the hydrogen evolution activity due to synergistic effects of alloy constituents.

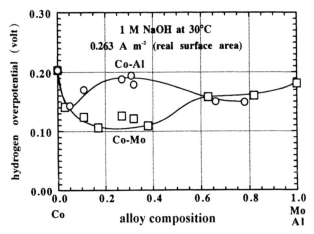

Fig. 6.7. Change in hydrogen overpotential of Co-Mo and Co-Al alloys estimated from polarization curves and real surface area [5]

The beneficial effect of molybdenum addition is more pronounced than that of aluminum addition. In order to obtain more active cobalt alloys, Co-Mo-Al alloys were prepared by sputter deposition. Figure 6.8 shows potentiodynamic polarization curves of these alloys. The Co-10Mo-58Al alloy shows the higher activity for hydrogen evolution. Its hydrogen overpotential at 10^3 A m^{-2} is more than 0.3 volts lower than that of cobalt metal. In general, the cobalt alloys containing a higher concentration of aluminum and lower concentration of molybdenum possess higher activity for hydrogen evolution. The examination of roughness factor revealed that the activity is in good agreement with the effective surface area. For example, the roughness factor

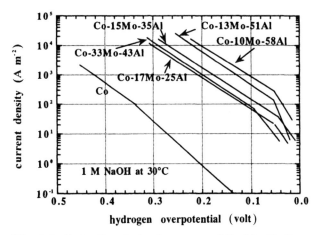

Fig. 6.8. Cathodic polarization curves of Co-Mo-Al alloys measured in a deaerated 1 M NaOH at 30°C [5]

of the Co-10Mo-58Al alloy was 5100, while that of binary Co-Mo or Co-Al alloy with the highest activity in each alloy family was about 1000. Consequently, both high surface area and synergistic effects of alloy constituents enhance the activity of these alloys for hydrogen evolution. In these studies sputter deposited Ni-15Mo alloy exhibits the highest activity for hydrogen evolution in the alkaline solution. It was, therefore decided to use Ni-15Mo alloy sputter-deposited on expanded nickel metal as the cathode for seawater electrolysis.

6.3.2 Anode Materials

The most difficult subject was to tailor anode materials for seawater electrolysis. In general, industrial seawater electrolysis is carried out for the production of chlorine. However, chlorine emissions into atmosphere are not allowed for preventing global warming. The equilibrium potential for oxygen evolution in neutral solutions is about 700 mV lower than that of chlorine evolution. However, oxygen evolution is a complicated reaction consisting of several elemental reactions while the chlorine evolution reaction is a simple reaction. Accordingly, the practical potential for seawater electrolysis readily exceeds the chlorine evolution potential, and chlorine evolution inevitably occurs.

Since manganese oxides have lower oxygen overpotentials, the enhancement of the oxygen evolution efficiency of manganese oxides was performed by adding various elements to manganese oxides. A simple method to prepare oxide electrodes containing two or more cations in a wide composition range is thermal decomposition of salt mixtures coated on the bulk substrate metals. In this examination buthanol or aqueous solutions of mixtures of 0.2 M $Mn(NO_3)_2$ and various metal salts were coated on a substrate. The substrates used were IrO_2-coated titanium. The presence of an intermediate IrO_2 layer between titanium and manganese oxide mixtures is necessary to prevent the formation of insulating titanium oxide between the electrocatalytically active manganese oxides and the titanium substrate during seawater electrolysis. After coating of the salt mixtures of manganese and other cations on the substrate thermal decomposition was carried out in air at 450°C. Coating and thermal decomposition were repeated until the formation of oxide mixtures of about 8×10^{-2} mol m^{-2}. The electrodes thus prepared are called (Mn-M)O_x/IrO_2/Ti electrodes. The electrode performance was examined by electrolysis of 0.5 M NaCl solution at 30°C. The oxygen evolution efficiency was estimated as the difference between the total charge passed and the charge for chlorine formation. The charge for chlorine formation during electrolysis was measured by iodimetric titration of chlorine and hypochlorite.

The oxygen evolution efficiency of the thermally decomposed manganese oxide electrode was 70%. Additions of various elements improved the oxygen evolution efficiency. Among them molybdenum [8] and tungsten [9] additions

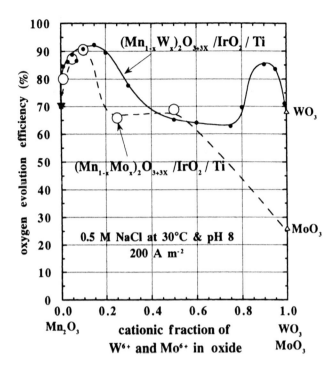

Fig. 6.9. Change in oxygen evolution efficiencies of $(Mn_{1-x}Mo_x)_2O_{3+3x}/IrO_2/Ti$ [8] and $(Mn_{1-x}W_x)_2O_{3+3x}/IrO_2/Ti$ [9] electrodes measured at 200 A m^{-2} in 0.5 M NaCl at 30°C and pH 8 with Mo^{6+} and W^{6+} contents of oxides

were particularly effective in decreasing chlorine evolution efficiency. Figure 6.9 shows oxygen evolution efficiencies of these oxide electrodes as a function of contents of additive elements. The addition of 1–10 mol.% molybdenum to the oxide leads to an increase in the oxygen evolution efficiency.

In particular, the electrode containing 10 mol.% molybdenum exhibits more than 90% oxygen evolution efficiency. However, the addition of an excess amount of molybdenum is detrimental. The tungsten addition is more interesting. WO$_3$ itself shows considerable oxygen evolution efficiency, namely 68%. Small additions of W^{6+} to Mn$_2$O$_3$ and small additions of Mn^{2+} to WO$_3$ are both effective in enhancing the oxygen evolution efficiency, although excess additions are ineffective.

Figure 6.10 shows X-ray diffraction patterns of thermally decomposed Mn-W oxides [9]. Addition of small amounts of tungsten results in the formation of nanocrystalline supersaturated solid solution oxides, while in the intermediate range a crystalline double oxide, MnWO$_4$, is formed. For Mn-Mo oxides small addition of molybdenum leads to supersaturation of molybdenum in Mn$_2$O$_3$-type oxides and the addition of 25 mol.% or more molybdenum gives rise to the formation of MnMoO$_4$.

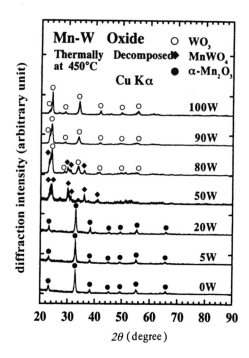

Fig. 6.10. X-ray diffraction patterns of manganese-tungsten oxide electrodes with different tungsten contents of oxides prepared by thermal decomposition at 450°C [9]

Accordingly, as far as supersaturated solid solutions are formed for Mn-Mo and Mn-W oxides, they show significant oxygen evolution efficiency, but the formation of double oxides, $MnWO_4$ and $MnMoO_4$ is not effective in enhancing the oxygen evolution efficiency. In this manner, additions of molybdenum and tungsten to manganese oxide are effective in enhancing the oxygen evolution efficiency in seawater electrolysis, but the maximum oxygen evolution efficiency for thermally decomposed oxides is about 90% and excess additions of these elements prove detrimental. Furthermore, manganese oxide formed on the IrO_2/Ti substrate by thermal decomposition is a trivalent manganese oxide such as the α-Mn_2O_3 type oxide, and such a lower oxide is apt to be dissolved by anodic oxidation during seawater electrolysis.

On the other hand, a tetravalent manganese oxide γ-MnO_2 can be formed by anodic deposition from divalent manganese salt solutions. As shown in the above experiments the addition of tungsten or molybdenum to manganese oxides is particularly effective in enhancing the oxygen evolution efficiency in 0.5 M NaCl solution. Since hexavalent tungsten and molybdenum dissolves in aqueous solutions in the form of WO_4^{2-} and MoO_4^{2-}, these anions should be drawn to the anodically polarized electrode during anodic deposition of manganese oxide. Accordingly, an attempt was made to prepare tungsten-

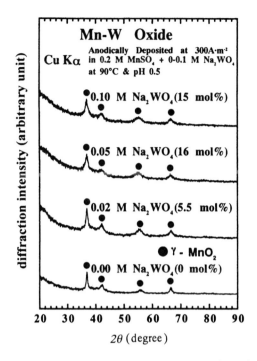

Fig. 6.11. X-ray diffraction patterns of anodically deposited manganese-tungsten oxides with different tungsten contents [10]

containing MnO_2 type oxide by anodic deposition on the IrO_2/Ti substrate. Electron probe microanalysis revealed that tungsten was included in the manganese oxide thus formed. Figure 6.11 shows X-ray diffraction patterns of manganese-tungsten oxides thus formed on the IrO_2/Ti substrate. Anodic deposition leads to the formation of γ-MnO_2-type oxides and the addition of tungsten results in nanocrystallization but does not form a new crystalline phase.

Figure 6.12 shows true polarization curves of anodically deposited MnO_2 and $(Mn_{0.84}W_{0.16})O_{2.16}$ electrodes for oxygen and chlorine evolution [10]. The $(Mn_{0.84}W_{0.16})O_{2.16}$ electrode exhibits higher oxygen evolution current and lower chlorine evolution current than MnO_2 electrode.

This indicates that tungsten addition increases the catalytic activity for oxygen evolution and decreases that for chlorine evolution. Figure 6.13 shows the oxygen evolution efficiency of the $(Mn_{0.84}W_{0.16})O_{2.16}$ electrode as a function of current density in electrolysis of 0.5 M NaCl solution [10]. A 99.6% oxygen evolution efficiency is seen at a current density of 200 A m^{-2}. Such a small amount of chlorine can be readily removed by the reaction with active charcoal, and consequently, we decided to use $(Mn_{1-x}W_x)O_{2+x}$ electrodes as the anode for seawater electrolysis. Recently, we succeeded in preparing Mn-Mo oxide electrodes showing 100% oxygen evolution efficiency [11].

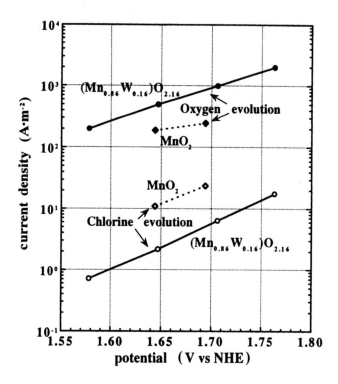

Fig. 6.12. True polarization curves of anodically deposited MnO_2 and $(Mn_{0.84}W_{0.16})O_{2.16}$ electrodes for oxygen and chlorine evolution obtained from galvanostatic polarization in 0.5 M NaCl at 30°C and pH 8 and from iodimetric titration of chlorine formed during polarization [10]

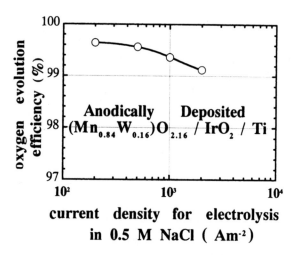

Fig. 6.13. Oxygen evolution efficiency of anodically deposited $(Mn_{0.84}W_{0.16})O_{2.16}$ on IrO_2-coated titanium substrate [10]

6.3.3 Catalysts for CO_2 Methanation

Effective catalysts for CO_2 methanation have been obtained from amorphous iron-group metal–valve–metal alloys. Under the reaction condition of equation (6.1), valve metals are oxidized but iron group elements are in the metallic state. This situation can readily be realized by oxidation-reduction pretreatment of these alloys. Figure 6.14 shows the rate of CO_2 conversion of Ni–valve–metal alloy catalysts [12]. The Ni-Zr alloy catalyst has exceptionally high catalytic activity for conversion of CO_2 at 1 atm. Three other nickel alloy catalysts have almost the same activity. The activity of iron–valve–metal alloy catalysts was more than three orders of magnitude lower than that of Ni-Zr alloy catalysts. A further important characteristic of the Ni-Zr alloy catalysts is their selectivity.

As shown in Figure 6.15 the conversion product on Ni-Zr alloy catalysts is almost 100% CH_4 and the very small amount of by-product is C_2H_6. In addition to the low activity of the iron–valve–metal alloy catalysts, the main product on them was carbon monoxide although small amounts of CH_4, C_2H_6, C_2H_4 and C_3H_8 were formed. Accordingly, in addition to their high activity, the Ni-Zr alloy catalysts can be regarded as ideal catalysts from the view point of selectivity, because the most important requirement in the conversion of CO_2 is to form currently used fuel.

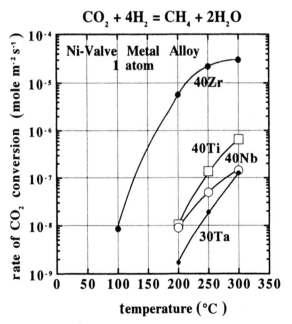

Fig. 6.14. Rates of CO_2 conversion on unit BET surface area of the catalysts prepared from Ni-valve metal alloys as a function of reaction temperature [12]

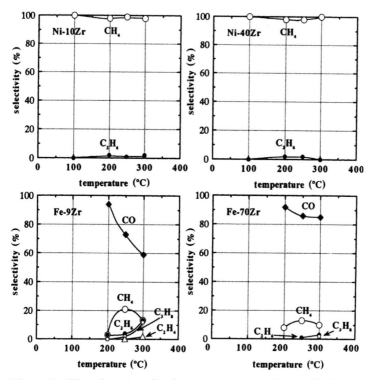

Fig. 6.15. The selectivity of carbon-containing products of hydrogenation of carbon dioxide on the catalysts prepared from amorphous Ni-Zr and Fe-Zr alloys [12]

If a new fuel is formed such as methanol, the development of combustion systems for the new fuel and its distribution to the world are necessary to treat CO_2 on a global scale, but those are very heavy burden for our society, and accordingly, such a conversion does not make sense as a means of combatting global warming.

Figure 6.16 shows a comparison of activities for the conversion of CO_2 to CH_4 on various catalysts [13]. The catalyst prepared from amorphous Ni-40Zr alloy shows particularly high activity. This catalyst consists of metallic nickel supported on nanocrystalline monoclinic and tetragonal ZrO_2. The tetragonal ZrO_2 is essentially metastable but is stabilized by inclusion of nickel ions. The catalytic activity of nickel supported on the tetragonal ZrO_2 is particularly high. This catalyst can be formed by the oxidation-reduction treatment of amorphous Ni-Zr alloys. The reaction shown in (6.1) forms H_2O as well as CH_4, and because of the establishment of equilibrium between reactants and products the maximum conversion by a single reactor is about 93 %. When a series of two reactors are used and when water is removed after passing the first reactor, the 99% conversion of CO_2 to CH_4 is realized as shown in Fig. 6.17 [13].

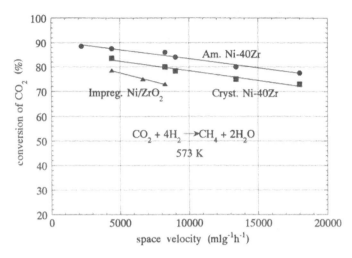

Fig. 6.16. Comparison of conversion of CO_2 to CH_4 by the reaction with H_2 on the catalyst prepared from amorphous Ni-40Zr alloy with those on catalyst prepared from crystalline Ni-40Zr alloy and on conventional Ni catalyst supported on ZrO_2 powder [13]

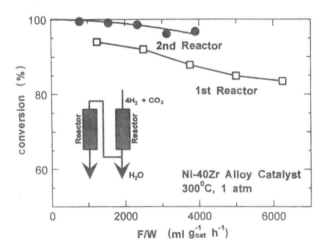

Fig. 6.17. Change in conversion from CO_2 to CH_4 by the reaction with H_2 on the catalyst prepared from amorphous Ni-40Zr alloy [13]. A series of double reactors were used and water was removed after passing the first reactor

6.4 A Global CO_2 Recycling Plant for Substantiation of the Idea

Because we succeeded in tailoring these key materials, we went to propose the scheme of global CO_2 recycling. Fortunately we were awarded special funding in 1996, and hence a prototype plant for global CO_2 recycling for substantiation of our idea was built on the roof top of the institute (IMR). A picture of it is shown in Figure 6.18. The plant connects a solar cell unit on a desert, 32 electrolytic cells for H_2 production and a series of two reactors for conversion of CO_2 to CH_4 by the reaction with H_2 at the coast close to the desert, and a CH_4 combustion and CO_2 recovery unit in an energy consuming district.

It has been proved that the energy consumer can really use solar energy from the desert in the form of CH_4 without emitting CO_2.

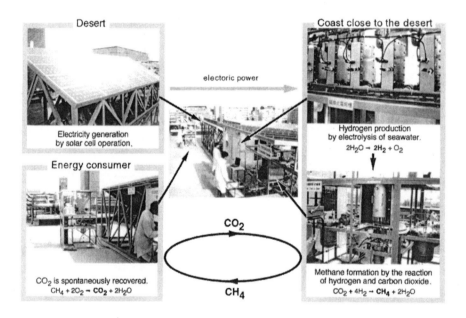

Fig. 6.18. The CO_2 recycling plant built on the roof of Institute for Materials Research, Tohoku University for substantiation of the idea of the global CO_2 recycling

6.5 Energy Balance and Amounts of Reduction of CO_2 Emissions

Since Egyptian politicians and scientists agreed to colaborate with us, the energy balance and the amount of reduction of CO_2 emissions in the global CO_2 recycling between the Middle East and Japan could be estimated for the operation of a 1 GW CH_4-combustion power plant in Japan [14]. The energy consumed in a year up to liquefaction of CH_4 including that corresponding to the construction of solar cell units and other units is almost the same as that spent up to obtaining LNG from a natural gas well. As shown in Figure 6.19 the energy necessary for the global CO_2 recycling is only 8.7% higher than the energy necessary for LNG combustion for power generation in Japan without control of CO_2 emissions. The extra 8.7% higher energies are for recovery, liquefaction and transportation of CO_2.

However, Egyptian scientists suggested to us that in an initial attempt, we do not need to carry recovered CO_2 from Japan to Egypt consuming energy and emitting CO_2 for transportation, but that CO_2 can be recovered in Egypt. If it is the case, the liquefied CH_4 production in the global CO_2 recycling is not very largely different from the production of LNG from the energetic point of view.

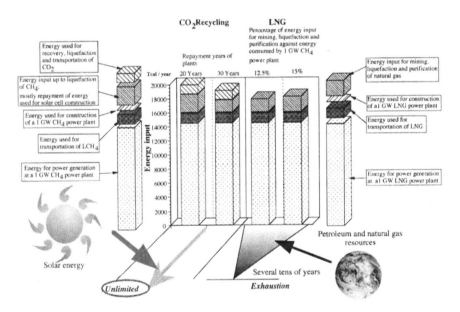

Fig. 6.19. A comparison of the energy input between the global CO_2 recycling and LNG combustion without CO_2 emission control in a 1 GW power plant for one year [14]

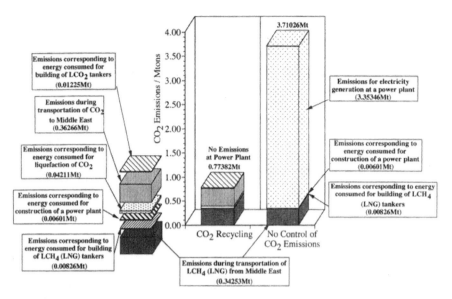

Fig. 6.20. A comparison of CO_2 emissions between the global CO_2 recycling and LNG combustion without CO_2 emission control in a 1 GW power plant for one year [14]

As shown in Fig. 6.20, the reduction of CO_2 emissions by global CO_2 recycling is 79% of CO_2 emissions from an LNG combustion power plant, that is, 2.62 Mtons/year. If transportation of recovered CO_2 from Japan to Egypt is unnecessary, the reduction of CO_2 emissions is almost 90%.

6.6 Economy of the Global CO_2 Reduction

The seawater electrolysis and CH_4 synthesis in the global CO_2 recycling do not require sophisticated advanced technology, since the seawater electrolysis is simple and since the CH_4 synthesis can be done at 1 atm. simply by passing a mixture of CO_2 and H_2 through our catalysts. Thus, the price of CH_4 produced in the global CO_2 recycling is mostly determined by the price of solar cells. According to the guideline given by the Japanese government, the price for solar cells installed at individual houses in Japan would become reasonable by technology improvements and mass production from 35 kW of accumulative amount of solar cells in 1995 to 4,600,000 kW in the near future. If this is realized, because the solar energy available on deserts in Egypt is more than 3 times higher than that in Japan due to longer sunlight time, and because of the mass-purchase of solar panels, the price of liquefied CH_4 produced by the global CO_2 recycling will become competitive with that of LNG.

6.7 Concluding Remarks

As shown in Fig. 6.1, CO_2 emissions in 2015 will be more than 1.6 times higher than in 1990. Accordingly, the amount of CO_2 which should be recovered in the very near future for prevention of global warming will be almost the same as the total amount of current CO_2 emissions in the whole world.

The global CO_2 recycling will be one of the most useful methods to prevent global warming and to supply abundant renewable energy. We are proposing a project for a pilot plant experiment of the global CO_2 recycling which will be carried out in Egypt in collaboration with Egyptian scientists for industrial design.

References

1. IN OUR HANDS, EARTH SUMMIT' 92, Asahi Shinbun, 1992, p.139
2. Energy Information Administration/International Energy Outlook 1997/Internet
3. K. Hashimoto: Mater. Sci. Eng. **A179/A180**, 27-30 (1994)
4. H. Kita: J. Electrochem. Soc. **113**, 1095-1106 (1966)
5. T. Aihara, A. Kawashima, E. Akiyama, H. Habazaki, K. Asami and K. Hashimoto: Mater. Trans. JIM **39**, 1017-1023 (1998)
6. A. Kawashima, E. Akiyama, H. Habazaki and K. Hashimoto: Mater. Sci. Eng. **A226-228**, 905-909 (1997)
7. A. Kawashima, T. Sakaki, H. Habazaki and K. Hashimoto: Mater. Sci. Eng. **A267**, 246-253 (1999)
8. K. Izumiya, E. Akiyama, H. Habazaki, N. Kumagai, A. Kawashima and K. Hashimoto: Mater. Trans. JIM, **38**, 899-905 (1997)
9. K. Izumiya, E. Akiyama, H. Habazaki, N. Kumagai, A. Kawashima and K. Hashimoto: Mater. Trans. JIM, **39**, 308-313 (1998)
10. K. Izumiya, E. Akiyama, H. Habazaki, N. Kumagai, A. Kawashima and K. Hashimoto: Electrochim. Acta **43**, 3303-3312 (1998)
11. K. Fujimura, T. Matsui, K. Izumiya, N. Kumagai, E. Akiyama, H. Habazaki, A. Kawashima, K. Asami and K. Hashimoto: Mater. Sci. Eng. **A267**, 254-259 (1999)
12. H. Habazaki, T. Tada, K. Wakuda, A. Kawashima, K. Asami and K. Hashimoto: in Corrosion, Electrochemistry and Catalysis of Metastable Metals and Intermetallics, C. R. Clayton and K. Hashimoto Eds., The Electrochemical Society, 1993, pp. 393-404
13. K. Shimamura, M. Komori, H. Habazaki, T. Yoshida, M. Yamasaki, E. Akiyama, A. Kawashima, K. Asami and K. Hashimoto: Supple. to Mater. Sci. Eng. **A226-228**, 376-379 (1997)
14. K. Hashimoto, E. Akiyama, H. Habazaki, A. Kawashima, M. Komori, K. Shimamura and N. Kumagai: Science Reports RITU A43[2], **659** 153-160 (1997)

7 Formation of Nano-sized Martensite and its Application to Fatigue Strengthening

M. Shimojo and Y. Higo

Precision and Intelligence Laboratory, Tokyo Institute of Technology, 4259 Nagatsuta-cho, Midori-ku, Yokohama, 226-8503, Japan.

Summary. In this chapter, the formation of micro and nano-sized martensite particles and their application to materials strengthening are described. The size of martensite can be controlled by the control of dislocation density and temperature. It is implied that the nucleation site of martensite is the intersection of two partial dislocations, however, further studies are necessary. Fatigue strength increased especially in the high cycle regime including fatigue limit due to the existence of nano-sized martensite particles.

7.1 Introduction

Martensitic transformation is a diffusionless transformation, which occurs in over-cooled metals and alloys. A number of investigations have been carried out on the martensitic transformation in steels. It has been known that martensitic transformation occurs heterogeneously, i.e., the nucleation of martensite occurs from dislocations, deformation bands, twin boundaries, grain boundaries or the surface of the material. However, the detail of the nucleation mechanism of martensite is still unclear. Martensite in steels usually grows at velocities approaching the speed of sound. It was thus difficult to observe the nucleation process and to control the size of martensites.

Martensites are known to have significant effects on the strength of steels. Considerable amount of research have been performed on the effects on the strength of steels. However, it has been known that using martensitic transformation was not very effective for fatigue strengthening in steels. This was probably due to the brittleness of the martensites generated in steels.

In this chapter, recent research in the formation of very small martensite particles in stainless steels and its application to fatigue strengthening is described.

7.2 Formation of Micro-sized Martensite

Austenitic steels (face-centred cubic structure, fcc) transform in part into a martensite which has a body-centred cubic (bcc) or body-centred tetragonal (bct) structure [1,2]. A great deal of research has been performed on crystalline structure of martensite [1,2], and crystallographic correspondence of martensite and its parent phase [3]. As the transformation proceeds in an auto-catalytic way, the propagation velocity of martensitic transformation is thought to be very fast in steels, though only a few attempts have been made to measure the velocity of martensitic transformation. Bunshah and Mehl [4] reported a velocity of 1000 m/s in an Fe-Ni alloy by means of electric resistivity measurements. Takashima [5] reported that the velocity was in a range between 110 and 200 m/s in a stainless steel using an acoustic emission technique. Thus, it was believed to be too difficult to control the size of small martensite in steels.

Recently, Borgenstam, et al. [6] measured M_g-M_s, where M_g and M_s are the temperatures below which martensite can grow if it has already nucleated and below which the formation of martensite starts upon cooling, respectively. They showed that the value was considered to be less than or equal to 8 K, which was a surprisingly low value. From their result, they proposed a hypothesis that martensitic transformation is controlled by growth rather than nucleation.

It could be possible to control the size of martensite, if the transformation is controlled by growth, as martensite growth should proceed at a finite speed. In this section, the formation of submicron-sized martensite is described.

The material employed was type 304 stainless steel rods (a pre-strain of 2%), which had been hot-extruded. The chemical composition of the steel is shown in Table 7.1. A transmission electron micrograph indicating dislocation structure and apparent density of the as-received specimen is shown in Fig. 7.1.

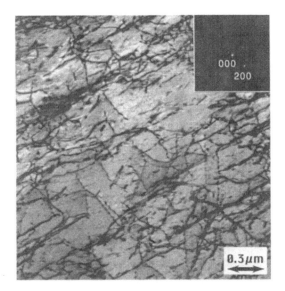

Fig. 7.1. Transmission electron micrograph indicating dislocation structure and apparent dislocation density of the as-received type 304 specimen

Table 7.1. Chemical compositions of type 304 stainless steel (mass%)

C	Si	Mn	P	S	Ni	Cr	Fe
0.06	0.31	1.31	0.034	0.028	8.11	18.41	BAL

The specimen was then cooled down to 195 K for two hours. The temperature to which the specimen was cooled down was determined by a preliminary test using another specimen as described below. The temperature at which burst-type martensitic transformation starts (M_b) was measured by detecting acoustic emission (AE) generated during the transformation when the specimen was cooled at a rate of approximately 1 K/min using liquid nitrogen. Acoustic emission voltage (root mean square) versus temperature for the type 304 specimen is shown in Fig. 7.2. Burst-type AE was detected at a temperature of 193 K. Metallographic observation revealed that relatively large martensite had already formed after the burst-type AE was detected. Thus, we determined the temperature to which the specimen was cooled down was 2 K higher than the M_b temperature, which was 195 K. (Note that M_b

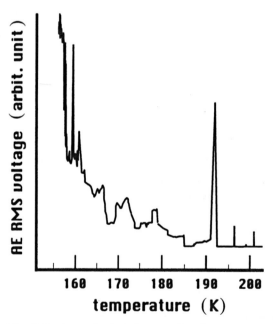

Fig. 7.2. Acoustic emission amplitude versus temperature for type 304 stainless steel. The specimen was cooled down at a rate of 1 K/min, showing that a burst-type martensitic transformation started at a temperature of 193 K

temperature greatly depends on chemical composition and the amount of pre-strain.)

Specimens before and after this cryogenic treatment were observed using a scanning laser microscope (SLM) and a transmission electron microscope (TEM).

Figure 7.3 shows specimen surface observation using the SLM of before and after the cryogenic treatment. No major difference was found on the photos. This indicates that no large martensite was formed after the treatment. A bright field TEM image of the treated specimen is shown in Fig. 7.4a. Dislocations and some small dots are observed. The size of these dots was

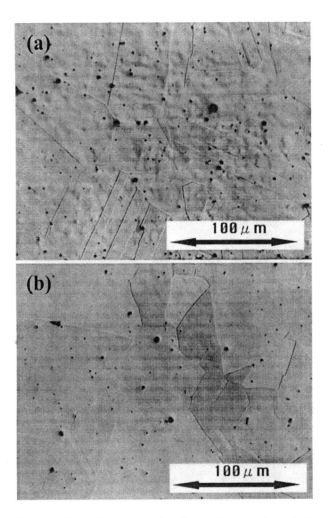

Fig. 7.3. Optical micrographs of type 304 specimen before (a) and after (b) the cryogenic treatment

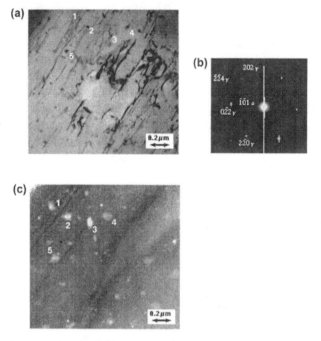

Fig. 7.4. Bright field transmission electron micrograph (a) of a cryogenic-treated specimen, the diffraction pattern of this image (b) showing both the parent phase (γ) and martensite (α) exist, and a dark field image from a bcc spot (c) showing micro-sized martensite particles are formed in type 304 specimen

approximately 20-100 nm in diameter. In order to identify the small dots, the diffraction pattern of this image was analysed. The result is shown Fig. 7.4b.

Two sets of diffraction patterns were observed. One is a diffraction pattern of austenite (γ) and the other is of a bcc structure (α). The orientation relationship between these is close to the Nishiyama relationship, which is $(111)_\gamma//(110)_\alpha$, $[112]_\gamma//[110]_\alpha$. Thus, it is considered that the bcc phase is martensite. A dark field image generated by a spot in the diffraction pattern of the martensite phase is shown in Fig. 7.4c. Some small bright dots were observed. Comparing the positions of dots in Fig. 7.4a and in Fig. 7.4c, it was confirmed that the positions of the dots coincide each other. It is therefore shown that submicron sized martensite particles are formed in the type 304 stainless steel.

The theories of martensite nucleation have been reviewed by Porter and Easterling [7]. It is shown in their book that martensite nucleation is a heterogeneous process and that surfaces and grain boundaries are not significantly contributing to nulceation. It is considered that martensite nucleation occurs at some defects within the crystals. The role of dislocation in martensite nucleation is also discussed in the book.

The characteristics of martensitic transformation occurred by cryogenic treatments is affected by the microstructure of the material such as dislocation density [8]. Martensitic transformation is also induced by plastic deformation. This is called deformation-induced or strain-induced martensitic transformation, when the material is plastically deformed below a certain temperature called M_d [1]. Plastic deformation in metals and alloys is in part a result of dislocation motion. Thus, some relation may exist between the cryogenic treatment-induced and the deformation-induced martensite transformations. Bogers and Burgers [9] showed that the transformation of the fcc lattice into bcc could be accounted by two directions of shear deformation. Olson and Cohen [10] extended this model to strain-induced transformation and concluded that the nucleation sites of α martensites were generated by two intersecting shear systems in austenite with the elements $\{111\}\langle 112\rangle_\gamma$.

The martensite variants formed by plastic deformation strongly depend on the direction of the stress applied [11]. In this study, the martensitic transformation is induced by the cryogenic treatment and not by plastic deformation. All of the dots in Fig. 7.4a do not correspond to the dots in Fig. 7.4c. This may indicate the existence of martensite of another variant which was not related to the analysed orientation.

Submicron-sized martensite particles have been formed in a type 304 stainless steel by controlling temperature to which the material was cooled down. The temperature was determined to be 2 K above the M_s temperature at which the "burst"-type transformation starts to occur.

7.3 Formation of Nano-sized Martensite

In the former section, the formation of submicron-sized martensite was described. According to Olson and Cohen [10], in a stainless steel with a low stacking fault energy, the nucleation sites of α embryos were placed at two intersecting shear systems in austenite. That is on the basis of the double shear mechanism proposed by Bogers and Burgers [9], in which a true b.c.c. structure is formed at the intersection of two shear bands which have particular Burgers vectors. As the intersection of two partial dislocations has a b.c.c.-like stacking, which is similar to the structure of α martensite, it is thought that the intersection of two partial dislocations can also be a nucleation site of a martensite.

As the growth of martensite requires plastic deformation of the parent phase, it may be expected that the size of martensite particles is controlled by the yield stress of the parent phase. In this section, the dislocation density, which changes the yield stress, is increased and an attempt to produce finer martensite particles are described. The materials employed were types 304 and 316 stainless steel rods, which had been hot-extruded. The type 304 stainless steel was the same one as used in the previous section. The chemical composition of the type 316 stainless steel is shown in Table 7.2. In the

Table 7.2. Chemical composition of type 316 stainless steel (mass %).

C	Si	Mn	P	S	Ni	Cr	Mo	Fe
0.05	0.34	1.35	0.031	0.027	10.10	16.86	2.05	BAL

previous section, the formation of micro-sized martensite particles in the type 304 stainless steel which had received a pre-strain of 2% was described. In this section, the pre-strain (i.e., dislocation density) was increased for the type 304 stainless steel in order to suppress the growth of martensite. A pre-strain of 10% was applied to the type 304 stainless steel using a tensile testing machine at a temperature of 473 K in order to increase dislocation density without forming strain-induced martensite. Specimens were prepared from the materials by machining. A TEM micrograph is shown in Fig. 7.5, indicating the increased dislocation density in the type 304 stainless steel. The specimens were then cooled down to 195 K for two hours for both the 304 and 316 specimens. The determination procedure of the temperature for type 304 stainless steel was written in the previous section. For the type 316 stainless steel, the same temperature was used though no burst-type AE was detected during the cooling down to 77 K, as shown in Fig. 7.6.

Specimens were thinned using a wheel cutter and emery paper. Electrochemical polishing was then performed using a perchloric acid and acetic acid

Fig. 7.5. TEM micrograph showing an increased dislocation density in 304 stainless steel after a pre-straining of 10%

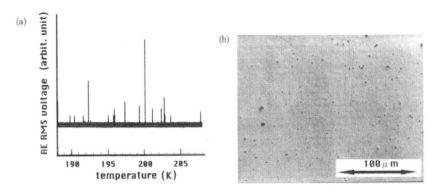

Fig. 7.6. Acoustic emission amplitude versus temperature for type 316 stainless steel (**a**) and a surface of the specimen cooled down to 77K (**b**), showing that no burst-type transformation occurred

solution to thin and eliminate the surface layer which might contain strain-induced martensite formed due to the cutting and grinding. The specimens were observed using a 200kV transmission electron microscope.

Figure 7.7 shows TEM micrographs of the type 304 specimen. Particles of approximately 5 nm in diameter are seen in the figure.

Spots from two phases appeared in the diffraction pattern, the angle of which seems to agree with the Nishiyama or the Kurdjumov-Sacks relationship. (It is unclear which of the relationships can be applied only from this diffraction pattern. However, the micro-sized particles had orientations close to the Nishiyama relationship, which is $(111)\gamma // (110)\alpha$, $[112]\gamma // [110]\alpha$. Thus, the existence of the Nishiyama relationship may be expected.) As indicated in the figure, a spot corresponds to $(111)\gamma$ and another corresponds to $(110)\alpha$. The dark field image using a spot from $(110)\alpha$ shows that the particles have a bcc structure. As the material was originally a single fcc phase and the new phase appeared after the cryogenic treatment, the particles are considered to be martensite. This indicates that the size of martensite can be controlled by a careful control of temperature and dislocation density. Figure 7.8 shows TEM micrographs of the type 316 specimen. Particles of approximately 5 nm in diameter, which are considered to be martensite, are seen in the figure.

Conventionally, martensitic transformation start temperature (M_s) has been defined as the temperature below which the martensitic phase is confirmed using optical microscopy in most research, including the paper written by Borgenstam, et al. [6]. However, this temperature is actually the temperature at which burst-like transformation starts, and we termed M_b in this chapter. According to the results in our research, $M_b < 77$ K for type 316 stainless steel. However, nano-sized martensite formed at 195 K. The temperature for martensite nucleation M_n is, therefore, much higher than the M_b temperature, if such a M_n temperature exists. As $M_b = 193$ K for type 304

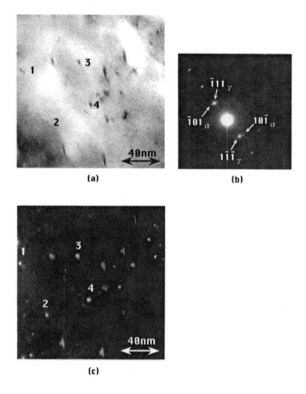

Fig. 7.7. Bright field transmission electron micrograph (**a**) of a cryogenic-treated specimen, the diffraction pattern of this image (**b**) showing both the parent phase (γ) and martensite (α) exist, and a dark field image from a bcc spot (**c**) showing nano-sized martensite particles are formed in type 304 specimen

stainless steel and very little growth occurred, if any, at 195 K, it is considered that M_g-M_b is less than 2 K. We can, therefore, hypothesise that $M_n \gg M_g > M_b$ and M_g-$M_b < 2$ K. Because M_g-M_b has a small value and it is not easy to observe nano-sized martensite, it is considered that both nucleation and burst were believed to occur at the same temperature, that is M_s.

A driving force for martensitic transformation is the free energy difference between fcc and bcc phases caused by over-cooling. The α martensitic transformation accompanies an increase in volume. This may cause plastic deformation of the parent phase. Thus, higher yield strength of the parent phase suppresses martensitic transformation or the growth of martensite particles. We, therefore, controlled both the temperature of the cryogenic treatment and yield strength, that is, dislocation density. (The yield strength of 10% pre-strained 304 is approximately 40% higher than that of the as-received material.) The formation of the nano-sized martensite may be the results of a fortuitous balance between the driving force which depends on temperature and the suppression of growth caused by the high dislocation

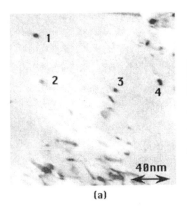
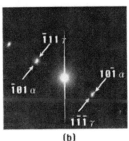

Fig. 7.8. Bright field transmission electron micrograph (**a**) of a cryogenic-treated specimen, the diffraction pattern of this image (**b**) showing both the parent phase (γ) and martensite (α) exist, and a dark field image from a bcc spot (**c**) showing nano-sized martensite particles are formed in type 316 specimen

density. In type 316 stainless steel, burst-type transformation is not likely to occur. It is, thus, considered that nano-sized martensite particles do not grow larger at the treatment temperature even in a relatively low dislocation density material.

Olson and Cohen [10] also described that the intersections of two ε martensite (hexagonal close-packed structure) plates or the intersections of ε martensite and active slip systems could be the nucleation sites. In the martensitic transformation induced by cryogenic treatment following pre-straining, as was done in this study, the possible nucleation sites are also considered to be the intersections of deformation bands which were produced by the prior pre-straining. However, the pre-straining was performed at a temperature of 473 K so that the formation of ε martensite was suppressed. Thus the major form of deformation was thought to be slip deformation, which is dislocation motion. It is therefore plausible that the intersections of dislocations or stacking faults may be the nucleation sites in this case. If there is a small martensite particle in an austenitic phase the misfit strain which is resulted in the difference in crystal structure would be very large and the particle may be unstable to exist. However, the intersection of partial dislocation has a bcc-like stacking, i.e., the dislocations absorb the misfit between the bcc particle and fcc matrix. This may be a reason for the existence of nano-sized martensite, though detailed studies are necessary on this matter. (No direct evidence of the nucleation at dislocation intersections was obtained because TEM observation of both intersecting dislocations and a martensite particle was not possible probably due to the large strain at the intersections.) If the martensite particles exist on the intersections of dislocations, dislocation motion should be suppressed once the martensite particles are produced by the cryogenic treatment. This can be a new type of strengthening, in which both work-hardening and precipitation hardening mechanisms are activated.

Nano-sized martensite particles have been formed in type 304 and 316 stainless steels by controlling temperature and dislocation density. This is considered to be the results of a fortuitous balance between the driving force which depends on temperature and the suppression of growth caused by the high dislocation density.

7.4 Application of Micro and Nano-sized Martensite to Materials Strengthening

Most of the strengthening methods for metallic materials, such as precipitation strengthening and solid solution strengthening, are supposed to be effective for monotonic loading. However, under cyclic loading these methods are not so effective as used under monotonic loading.

Prior to the onset of precipitation strengthening, some amount of dislocation motion is required, or dislocations must move until they encounter obstacles. Though any obstacle may stop dislocations, they can move in the reverse direction under the reversed stress of cyclic loading. Such a small amount of cyclic dislocation motion over many cycles may be sufficient to cause fatigue crack initiation. Once a crack has initiated, a reduction in ductility which generally occurs in strengthened metals results in a degradation of crack growth resistance.

In the previous sections, the formation of submicron and nano-sized martensite particles were described. Since fine martensitic particles are supposed to precipitate on intersections of dislocations, the particles are expected to suppress dislocation motion under cyclic loading. In this section, an application of small martensite particles to fatigue strengthening is described.

The materials used were two austenitic stainless steels, type 304 and type 316. Specimens were machined from the materials after dislocation density was controlled for the 304 stainless steel by pre-straining. No pre-straining was performed for the 316 stainless steel, but the dislocation density of the subsurface region of the specimens may be increased by machining, in this case, using a lathe. Martensite particles of approximately 5 nm were formed at intervals of approximately 40 nm in both the specimens by the cryogenic treatment written above. Tensile tests were performed using the specimen shown in Fig. 7.9a and an Instron-type tensile testing machine at room temperature at a cross head speed of 8×10^{-4} s^{-1}. Tension-compression tests to obtain the cyclic stress strain response of the material were performed using the specimens shown in Fig. 7.9b. A strain gauge was attached on each side of the specimen. The measurements were carried out using a servo-hydraulic fatigue machine at a cyclic frequency of 1 Hz at room temperature.

Fatigue life tests were performed using a rotary four-point bend fatigue machine at an R-ratio of -1 in laboratory air. The dimensions of the fatigue specimen is shown in Fig. 7.9c. The temperature of the specimen surface was measured using an infrared temperature sensor. Test frequency (revolution speed) was chosen to be set at 2-10 Hz so that the temperature of the specimen surface did not exceed 313 K during the test. (Heating of the specimen may occur due to cyclic plastic deformation.)

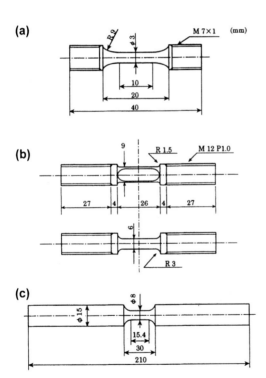

Fig. 7.9. Dimensions of specimens used for tensile test (**a**), cyclic stress-strain test (**b**) and fatigue life test (**c**)

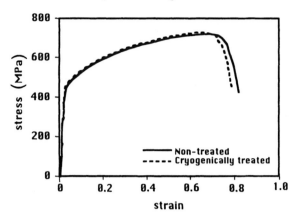

Fig. 7.10. Monotonic stress-strain curves of cryogenically treated and non-treated type 304 stainless steel

Stress-strain curves for monotonic loading on both cryogenically treated and non-treated specimens for the type 304 stainless steel (10% pre-strained) are shown in Fig. 7.10. The tensile properties of the treated specimens were almost the same as those of the non-treated specimen in all respects such as yield stress, tensile strength and elongation. It is considered that ductility was not reduced by the treatment.

Figure 7.11 shows plastic strain amplitude, obtained from the cyclic stress-

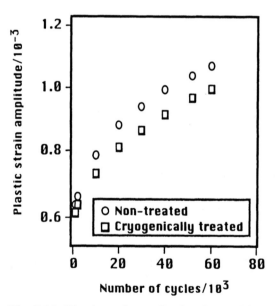

Fig. 7.11. Plastic strain amplitude, obtained from cyclic stress-strain response, as a function of the number of cycles under constant stress amplitude loading on type 304 stainless steel

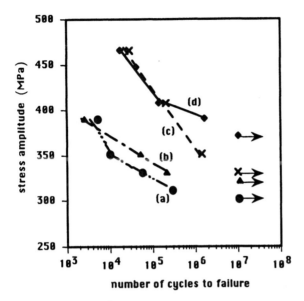

(a) as received
(b) micro-sized martensite dispersed
(c) cold-worked
(d) cold-worked and nano-sized martensite dispersed

Fig. 7.12. Fatigue properties of type 304 stainless steel

strain response, as a function of the number of cycles, during tests performed on the type 304 steel at a constant stress amplitude of 320 MPa. As the specimen was pre-strained and work hardened, plastic strain amplitude increased with the number of cycles. (This means that cyclic softening occurred in these materials.) The rate of cyclic softening is lower for the treated specimens (having nano-sized martensite particles) than for the non-treated specimens.

$S - N$ curves of the type 304 stainless steel are shown in Fig. 7.12. Fatigue life curves of the as-received and micro-sized martensite dispersed specimens used in a previous study [12] are also plotted in the figure. Fatigue lives of the nano-sized martensite dispersed specimens increased, especially in the high cycle regime (more than 10^6 cycles), as compared to those of the other specimens. The fatigue limit of the treated specimens (containing nano-sized martensite) is 370 MPa and is approximately 12% higher than that of the non-treated (cold-worked only) specimens. (Fatigue limit was defined, in this research, as the maximum stress amplitude at which the specimen does not fail until 10^7 cycles.) As the fatigue limit of the as-received specimen was 300 MPa, the combination of the pre-straining and cryogenic treatment increased the fatigue limit by more than 23%. However, no significant increase is seen in the low cycle regime (less than 10^5 cycles).

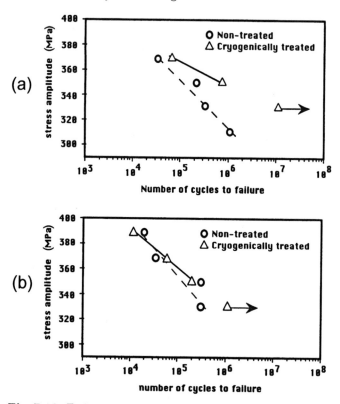

Fig. 7.13. Fatigue properties of type 316 stainless steel (a), and surface polished specimens (b) of the same material

Fatigue life curves of the type 316 stainless steel are shown in Fig. 7.13a. A similar trend to those of the type 304 stainless steel is found. No pre-straining was performed on the type 316 specimens, but the dislocation density in the subsurface regions of the specimens may be increased by machining.

Fatigue life curves of the type 316 specimens, the surfaces of which are polished using emery papers, are shown in Fig. 7.13b. No significant increase in fatigue life is found on the treated specimens compared with non-treated specimens. It is, therefore, considered that the nano-martensite particles formed in the subsurface region, which has a high dislocation density, increased the fatigue life of this material. (The effect of subsurface regions should be larger than that of the interior region because the fatigue life tests were carried out by rotary bending.)

Fracture surfaces of specimens tested at a stress of 331 MPa for micro-sized martensite containing specimen and as-received specimen observed using a SEM are shown in Fig. 7.14. No difference was observed on the fracture surfaces of both the specimens. This indicates that once cracks were initiated, the plastic strain at the crack tips were so large due to the stress concentra-

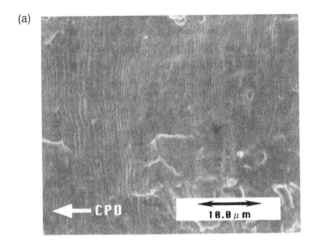

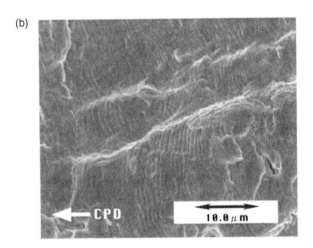

Fig. 7.14. (a) Fractographs of non-treated and treated (micro-sized martensite) (b) specimens of type 304 stainless steel, showing no difference in crack propagation behaviour

tions that the dislocation locking mechanism proposed in this study no longer operates.

The only difference between the treated and the non-treated specimens is the cryogenic treatment. TEM study revealed that small martensite particles were formed in the treated specimens. It is, therefore, suggested that the differences in fatigue behaviour come from the existence of small martensite particles.

The results of cyclic stress-strain tests indicate that cyclic plastic deformation is suppressed in the treated specimens. This means that cyclic dislocation motion may be suppressed due to the existence of nano-sized martensite particles. This may be the reason for the increase in fatigue life, though it is difficult to compare the cyclic stress-strain results with the fatigue life results directly, because the stress condition for the cyclic stress-strain tests was different from that for the fatigue life tests.

It has been reported that plastically induced ε martensitic transformation increases significantly the elongation to failure in tensile tests. This is not, however, the reason for the fatigue life extension, because this transformation could occur in both treated and non-treated specimens and this transformation could occur more readily at higher stress amplitudes. To investigate the effect of ε martensite on fatigue life, fatigue life measurements were carried out on a manganese steel (Hadfield steel). The $M_{s\varepsilon}$ was measured by means of the AE technique and it was found that ε martensite appears at 215 K in this steel. The microstructure after the cooling treatment is shown in Fig. 7.15. It has been confirmed that only ε martensite appears in this manganese steel when it is cooled down. Fatigue lives of this material are shown in Fig. 7.16. Fatigue life did not increase (rather decreased) for the cooled manganese steel compared with non-treated manganese steel. It is thus concluded that the formation of ε martensite adversely affects the fatigue life of steels.

In conventional martensite hardened materials, the martensite particles are relatively large and brittle so that the particles may act as crack initiation sites [13]. However in this nano-martensite dispersed material, it is thought that dislocations can come off the martensite particles at higher stresses because the particles are so small that the pinning force is small. This may be the reason for the reduced increase in fatigue life at higher stress amplitudes. This agrees with the fact that the tensile properties are almost the same for both the specimens. This is not necessarily a deficiency of this strengthen-

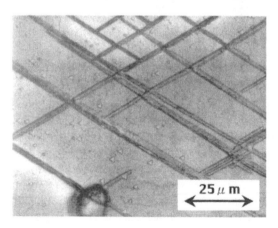

Fig. 7.15. The microstructure of manganese steel after cooling treatment at 215 K, showing ε martensites appeared

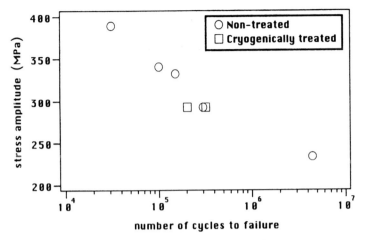

Fig. 7.16. Fatigue properties of manganese steel

ing method. Dislocation coming off at a high stress inhibits any decrease in ductility, which usually accompanies strengthening.

In conventional strengthening methods such as precipitation hardening, dislocations move until the dislocations encounter precipitates. Such hardening methods are not very effective for suppressing cyclic dislocation motion, which occurs under cyclic loading. However, the nano-sized martensite particles, which are considered to exist at the intersection of two partial dislocations, may restrict dislocation motion or make dislocation motion more reversible. As dislocations are fixed at their intersections, bowing motion could be possible, but the rearrangement of dislocation structures, which usually occurs in work-hardened materials, may be retarded. Detail of such mechanisms should be clarified in future work.

If materials are just cooled down to an appropriate temperature to form nano-sized martensite particles, both work-hardening and precipitation hardening mechanisms can be activated. This could be a new type of strengthening method, which is effective for extending fatigue life especially in the high cycle regime.

Fatigue strength increased especially in the high cycle regime including fatigue limit and no decrease in ductility was observed as compared to that in the stainless steels in which no particles were formed. This is probably due to the suppression of cyclic dislocation motion by the nano-sized martensite particles, which are considered to exist at intersections of dislocations. In this material, both work-hardening and precipitation hardening mechanisms can be activated. This could be a new type of strengthening method, which is effective for extending fatigue life especially in high cycle regime.

7.5 Conclusions and Future Work

In this chapter, the formation of micro and nano-sized martensite particles and their application to materials strengthening are described. The size of martensite can be controlled by the control of dislocation density and temperature. It is implied that the nucleation site of martensite is the intersection of two partial dislocations, however, further studies are necessary. Fatigue strength increased especially in the high cycle regime including fatigue limit due to the existence of nano-sized martensite particles.

References

1. Z. Nishiyama: Martensitic Transformation, Academic Press, New York, 1978.
2. W. D. Callister Jr.: Materials Science and Engineering - An Introduction -, 3rd ed., John Wiley & Sons Inc., New York, 1994, pp. 298-300.
3. P. L. Mangonon Jr and G. Thomas: Metall. Trans., **1**, 1577 (1970).
4. R. F. Bunshah and R. F. Mehl: Trans. AIME, **197**, 1251 (1953).
5. K. Takashima, Y. Higo and S. Nunomura: Phil. Mag. A, **49**, 231 (1984).
6. A.Borgenstam, M.Hillert and J. Agren: Acta Metall. Mater., **43**, 945 (1995).
7. D. A. Porter and K. E. Easterling: Phase Transformations in Metals and Alloys, Van Nostrand Reinhold Co. Ltd., Berkshire, UK., (1981), pp. 397-409.
8. Y. Higo, T. Mori and T. Nakamura: J. Iron Steel Inst. Jpn., **61**, 2561 (1975) (in Japanese).
9. A. J. Bogers and W. G. Burgers: Acta Metall., **12**, 255 (1964).
10. G. B. Olson and M. Cohen: J. Less Comm. Metals, **28**, 107 (1972).
11. Y. Higo, F. Lecroisey and T. Mori: Acta Metall., **22**, 313 (1974).
12. T. H. Myeong, Y. Yamabayashi, M. Shimojo and Y. Higo: Int. J. Fatigue, **19**, S69 (1997).
13. G. Franke and C. Altstetter: Metall. Trans. A., **7A**, 1719 (1976).

Some of the figures are reprinted from Int. J. Fatigue, vol. 19, T. H. Myeong, et al., A New Life Extension Method for High Cycle Fatigue using Micro-Martensitic Transformation in an Austenitic Stainless Steel, pp. S69-S73, Copyright (1997), with permission from Elsevier Science.

Index

activation energy 33, 52
amorphous alloy 1, 3–5, 17–20, 22, 24, 27, 28, 45, 71, 88, 97, 109, 114–116, 123, 169
amorphous material 87
amorphous phase 3, 12, 14, 15, 44, 64, 71, 75, 82, 87, 88, 90–94, 96, 101, 115, 133, 135, 137, 148, 157
amorphous structure 70, 101, 110, 122
anodic deposition 176
anomalous X-ray scattering 6, 20
atomic structure 107, 113

ball milling 87, 88, 95, 96
bulk amorphous alloy 11, 13, 25, 27, 34, 36, 43
Burgers vector 191

castability 48
catalytic activity 122, 125, 172, 177
cathodic polarization 122
chlorine evolution potential 174
CO_2 recycling 167
configurational entropy 66
coordination number 6
corrosion 125
corrosion resistance 38, 109
creep 26
critical cooling rate 4, 70
cryogenic treatment 191
crystalline phase 87, 92
crystallization 98, 101
crystallization behavior 82
crystallization temperature 3, 69, 90, 92, 93, 109, 114, 115, 142
Curie temperature 35
cyclic loading 196
cyclic transformation 93, 95

deformation behavior 40
density 44
diffusivity 52

electrical resistivity 141
electrocatalytic activity 116
electrochemical impedance 126
electrodeposition 87, 96, 97, 104
electron transport 70
energy-dispersive X-ray spectroscopy 97
enthalpy of transformation 150
extrusion 11, 44

fatigue strength 203
ferromagnetism 34
free energy 87

glass structure 84
glass transition 3, 18–20, 32, 64, 69–72, 75, 79, 80, 82, 83
glass-forming ability 3, 4, 13, 15, 38, 136, 140
global warming 115, 180
grain boundary 105
grain boundary thickness 106, 107
grain size 106, 108

Hall coefficient 83
Hall-Petch strengthening 105
heat of mixing 5, 6, 140
heterogeneous nucleation 10, 11
high frequency permeability 48
high resolution electron microscopy 97
high strain-rate superplasticity 41
homogeneous nucleation 5, 10
hot pressing 11, 147

hydrogen absorption   69, 84
hydrogen electrode   116
hydrogen evolution   116, 172
hydrogen overpotential   120, 125, 172

impact fracture   27
interference function   5

Kissinger plot   33

liquid quenching   110
local structure   90

magnetic properties   35, 37, 38, 92
magnetization   92, 93, 125
martensite   186–188, 193
mechanical alloying   88, 92, 93, 116, 127
mechanical properties   29, 61, 100, 114, 115
melt spinning   71, 133
metastable materials   87
microformability   39
mold casting   24
molecular dynamics   107

nano-martensite   200, 202
nanocrystalline   90, 93, 102, 105, 108, 116
nanocrystallite   107
nanocrystallization   101, 177
nanocrystallized alloy   97
nanostructured material   150
nucleation   7
nucleation and growth   3, 8, 10

oxygen evolution   177

partial dislocation   191
phase separation   34
phase transformation   88–90, 92–96, 108

phase transition   70
polarization curve   123, 173
potentiostatic polarization   126
powder consolidation   43

radial distribution function   6
rapid quenching   109
rapid solidification   87
reduced glass transition temperature   4
rod milling   88

shear band   26, 60
shear deformation   156
short-range order   19, 88
solar cell   115
solid state reaction   92
specific heat   19
spinodal decomposition   135
sputter deposition   116, 117, 122
sputtering   87, 109
stress overshoot   60
stress relaxation   52, 53, 58
stress-strain curve   56
structural analysis   97
structural changes   89, 96
structural relaxation   17, 19, 53
supercooled liquid   3–6, 8, 10, 11, 13, 15, 17–20, 23, 34, 39, 52, 69, 70, 75, 79, 82, 84
superplasticity   40
supersaturated solid solution   176

tensile strength   43, 198
tensile test   53
thermal stability   35, 39, 100
transition metal   116, 137

vickers microhardness   98

wave vector   36
work-hardening   26

Printing: Druckhaus Beltz, Hemsbach
Binding: Buchbinderei Schäffer, Grünstadt

LaVergne, TN USA
18 December 2009
167516LV00004B/3/P